GENERAL STRUCTURES
QUESTIONS & ANSWERS

John F. Hardt, AIA, Consulting Editor

KAPLAN) AEC EDUCATION

This publication is designed to provide accurate and authoritative information in regard to the subject matter covered. It is sold with the understanding that the publisher is not engaged in rendering legal, accounting, or other professional service. If legal advice or other expert assistance is required, the services of a competent professional person should be sought.

President: Roy Lipner
Vice-President of Product Development and Publishing: Evan M. Butterfield
Editorial Project Manager: Jason Mitchell
Director of Production: Daniel Frey
Quality Assurance Editor: David Shaw
Creative Director: Lucy Jenkins

Published by Kaplan AEC Education
a division of Dearborn Financial Publishing, Inc.®
A Kaplan Professional Company
30 South Wacker Drive, Suite 2500
Chicago, IL 60606-7481
(312) 836-4400
http://www.kaplanAECarchitecture.com

Printed in the United States of America.

06 07 08   10 9 8 7 6 5 4 3 2 1

CONTENTS

ARE OVERVIEW

Since the State of Illinois first pioneered the practice of licensing architects in 1897, architectural licensing has been increasingly adopted as a means to protect the public health, safety, and welfare. Today, all U.S. states and Canadian provinces require licensing for individuals practicing architecture. Licensing requirements vary by jurisdiction; however, the minimum requirements are uniform and in all cases include passing the Architect Registration Exam (ARE). This makes the ARE a required rite of passage for all those entering the profession, and you should be congratulated on undertaking this challenging endeavor.

Developed by the National Council of Architectural Registration Boards (NCARB), the ARE is the only exam by which architecture candidates can become registered in the United States or Canada. The ARE assesses candidates' knowledge, skills, and abilities in nine different areas of professional practice, including a candidate's competency in decision making and knowledge of various areas of the profession. The exam also tests competence in fulfilling an architect's responsibilities and in coordinating the activities of others while working with a team of design and construction specialists. In all jurisdictions, candidates must pass the nine divisions of the exam to become registered.

The ARE is designed and prepared by architects, making it a practice-based exam. It is generally not a test of academic knowledge, but rather a means to test decision-making ability as it relates to the responsibilities of the architectural profession. For example, the exam does not expect candidates to memorize specific details of the building code, but requires them to understand a model code's general requirements, scope, and purpose, and to know the architect's responsibilities related to that code. As such, there is no substitute for a well-rounded internship to help prepare for the ARE.

Exam Format

The ARE consists of nine divisions: three graphic and six multiple-choice. The multiple-choice divisions are timed and contain differing numbers of questions. The number of questions and time limit for each division is outlined in the table below. For detailed information on the graphic divisions, refer to the study guides for those divisions.

Division	Questions	Hours
Pre-Design	105	2.5
General Structures	85	2.5
Lateral Forces	75	2
Mechanical and Electrical Systems	105	2
Building Design/ Materials & Methods	105	2
Construction Documents & Services	115	3

The exam presents questions individually. Candidates may answer questions, skip questions, or mark questions for further review. Candidates may also move backward or forward within the exam using simple on-screen icons. The technique of marking questions for further review is a useful tool and is described in more detail below.

Actual appointment times for taking the exam are slightly longer than the actual exam time, allowing candidates to check in and out of the testing center. All ARE candidates are encouraged to review NCARB's *ARE Guidelines* for further detail about the exam format. These guidelines are available via free download at NCARB's Web site (*www.ncarb.org*).

ARCHITECTURAL HISTORY

Questions pertaining to the history of architecture appear in all of the multiple-choice divisions. The prominence of historical questions will vary not only by division but also within different versions of the exam for each division. In general, however, history tends to be lightly tested, with approximately three to seven history questions per division, depending upon the total number of questions within the division. One aspect common to all the divisions is that whatever history questions are presented will be related to that division's subject matter. For example, a question regarding Chicago's John Hancock Center and the purpose of its unique exterior cross bracing may appear on the Lateral Forces exam.

Though it is difficult to predict how essential your knowledge of architectural history will be to passing any of the multiple-choice divisions, it is recommended that you refer to a primer in this field—such as Kaplan's *Architectural History*—before taking each exam, and that you keep an eye out for topics relevant to the division for which you are studying. It is always better to be overprepared than taken by surprise at the testing center.

Question Format

It is important for exam candidates to familiarize themselves not only with exam content, but also with question format. Familiarity with the basic question types found in the ARE will reduce confusion, save time, and help you pass the exam. The multiple-choice divisions contain three basic question types.

The first and most common type is a straightforward multiple-choice question followed by four choices (A, B, C, and D). Candidates are expected to select the correct answer. An example of this type of question is shown below.

> Which of the following cities is the capital of the United States?
> A. New York
> **B. Washington, DC**
> C. Chicago
> D. Los Angeles

The second type of question is a negatively worded question. In questions such as this, the negative wording is usually highlighted using all caps, as shown below.

> Which of the following cities is NOT located on the west coast of the United States?
> A. Los Angeles
> B. San Diego
> C. San Francisco
> **D. New York**

The third type of question is a combination question. In a combination question, more than one choice may be correct; candidates must select from combinations of potentially correct choices. An example of a combination question is shown on the next page.

NEW TO THE EXAM

ARE 3.1

In November 2005 NCARB released *ARE Guidelines* Version 3.1, which outlines changes to the exam that are effective as of February 2006. These new guidelines primarily detail changes for the Site Planning division, which now combines the site design and site parking vignettes as well as the site zoning and site analysis vignettes. For more details about these changes, please refer to the study guides for the graphic divisions.

The new guidelines mean less to those preparing for multiple-choice divisions. Noteworthy points are outlined below.

- All division statements and content area descriptions are unchanged for the multiple-choice divisions.
- The number of questions and time limits for all exams are unchanged.
- The list of codes and standards candidates should familiarize themselves with has been reduced to those of the International Code Council (ICC), the National Fire Protection Association (NFPA), and the National Research Council of Canada.
- A statics title has been removed from the reference list for General Structures.

Rolling Clock

A rolling clock went into effect January 1, 2006. Candidates must now pass all nine ARE divisions within a five-year period. Additionally, NCARB has instituted a set of "transitional rules" for candidates already in the process of taking the ARE when the clock went into effect. See the new guidelines or visit the NCARB Web site for more detailed information.

Which of the following cities are located within the United States?

    I. New York

    II. Toronto

    III. Montreal

    IV. Los Angeles

A. I only

B. I and II

C. II and III

D. I and IV

Recommendations on Exam Division Order

NCARB allows candidates to choose the order in which they take the exams, and the choice is an important one. While only you know what works best for you, the following are some general considerations that many have found to be beneficial:

1. The Building Design/Materials & Methods and Pre-Design divisions are perhaps the broadest of all the divisions. Although this can make them among the most intimidating, taking these divisions early in the process will give a candidate a broad base of knowledge and may prove helpful in preparing for

subsequent divisions. An alternative to this approach is to take these two divisions last, since you will already be familiar with much of their content. This latter approach likely is most beneficial when you take the exam divisions in fairly rapid succession so that details learned while studying for earlier divisions will still be fresh in your mind.

2. The Construction Documents & Services exam covers a broad range of subjects, dealing primarily with the architect's role and responsibilities within the building design and construction team. Because these subjects serve as one of the core foundations of the ARE, it may be advisable to take this division early in the process, as knowledge gained preparing for this exam can help in subsequent divisions.

3. The General Structures and Lateral Forces divisions cover related and overlapping subjects. Take them consecutively, and take General Structures first, since it is broader and addresses fundamental principles necessary for success in Lateral Forces.

4. The three graphic divisions all use an identical software platform and employ similar graphic drawing tools. Because becoming fluent with this software is crucial to passing these exams, take the three graphic divisions sequentially.

5. The Mechanical & Electrical Systems and Building Technology exams cover loosely related material. As such, it is often beneficial to take these two exams consecutively.

6. Take exams that particularly concern you early in the process. NCARB rules prohibit retaking an exam for six months. Therefore, failing an exam early in the process will allow the candidate to use the waiting period to prepare for and take other exams.

EXAM PREPARATION

Overview

There is little argument that preparation is key to passing the ARE. With this in mind, Kaplan has developed a complete learning system for each exam division, including study guides, question-and-answer handbooks, mock exams, and flash cards. The study guides offer a condensed course of study and will best prepare you for the exam when utilized along with the other tools in the learning system. The system is designed to provide you with the general background necessary to pass the exam and to provide an indication of specific content areas that demand additional attention.

In addition to the Kaplan learning system, materials from industry-standard documents may prove useful for the various divisions. Several of these sources are noted in the "Supplementary Study Materials" section below.

Understanding the Field

The subject of structures may fall under the direct responsibility of the architect or under the responsibility of a structural consultant, depending on the scale and complexity of the project. In either case, however, properly designed building structures are critical in protecting the safety of building occupants. This significant role means that the subject of structural systems must be thoroughly understood in order for architects to properly integrate structural elements into their designs and permit constructive interaction with other members of the building design team.

Understanding the Exam

The general structures and lateral forces exams are among the most daunting of the ARE exams due to the technical nature of the subject and the wide array of problems that may be presented. Candidates will find questions covering statics, wood construction, concrete, steel, foundations, long-span structural systems, and connections frequently on the exam. Many candidates, however, allow the breadth of the exam to intimidate unnecessarily.

Candidates must be familiar with the *Manual of Steel Construction* published by the American Institute of Steel Construction, but it is not necessary to memorize any part of this volume. Rather, candidates should be aware of how to utilize the charts and tables found in the book. Tables and charts needed to solve problems within the ARE will be provided during the exam. Additionally, candidates should be aware that a series of equations, tables, and formulas are available within the ARE software by clicking on the References button. Therefore, candidates should not waste time memorizing formulas, but rather should focus their studies on practicing their use.

Many have also found that the question-and-answer books and mock exams published by Kaplan are especially useful for this exam as they provide additional opportunities to practice working through structural problems.

Preparation Basics

The first step in preparation should be a review of the exam specifications and reference materials published by NCARB. These statements are available for each of the nine ARE divisions to serve as a guide for preparing for the exam. Download these statements and familiarize yourself with their content. This will help you focus your attention on the subjects on which the exam focuses.

Though no two people will have exactly the same ARE experience, the following are recommended best practices to adopt in your studies and should serve as a guide.

Set aside scheduled study time.
Establish a routine and adopt study strategies that reflect your strengths and mirror your approach in other successful academic pursuits. Most importantly, set aside a definite amount of study time each week, just as if you were taking a lecture course, and carefully read all of the material.

Take—and retake—quizzes.
After studying each lesson in the study guide, take the quiz found at its conclusion. The quiz questions are intended to be straightforward and objective. Answers and explanations can be found at the back of the book. If you answer a question incorrectly, see if you can determine why the correct answer is correct before reading the explanation. Retake the quiz until you answer every question correctly and understand why the correct answers are correct.

Identify areas for improvement.
The quizzes allow you the opportunity to pinpoint areas where you need improvement. Reread and take note of the sections that cover these areas and seek additional information from other sources. Use the question-and-answer handbook and CD-ROM test bank as a final tune-up for the exam.

Take the final exam.
A final exam designed to simulate the ARE follows the last lesson of each study guide. Answers and explanations can be found on the pages following the exam. As with the lesson quizzes, retake the final exam until you answer every question correctly and understand why the correct answers are correct.

Use the flash cards.

If you've purchased the flash cards, go through them once and set aside any terms you know at first glance. Take the rest to work, reviewing them on the train, over lunch, or before bed. Remove cards as you become familiar with their terms until you know all the terms. Review all the cards a final time before taking the exam.

Supplementary Study Materials

In addition to the Kaplan learning system, materials from industry-standard sources may prove useful in your studies. Candidates should consult the list of exam references in the NCARB guidelines for the council's recommendations and pay particular attention to the following publications, which are essential to successfully completing this exam:

- International Code Council (ICC) *International Building Code*
- American Institute of Steel Construction *Manual of Steel Construction: Allowable Stress Design,* Ninth Edition

Test-Taking Advice

Preparation for the exam should include a review of successful test-taking procedures—especially for those who have been out of the classroom for some time. Following is advice to aid in your success.

Pace yourself.

Each division allows candidates at least one minute per question. You should be able to comfortably read and reread each question and fully understand what is being asked before answering.

Read carefully.

Begin each question by reading it carefully and fully reviewing the choices, eliminating those that are obviously incorrect. Interpret language literally, and keep an eye out for negatively worded questions.

Guess.

All unanswered questions are considered incorrect, so answer every question. If you are unsure of the correct answer, select your best guess and/or mark the question for later review. If you continue to be unsure of the answer after returning the question a second time, it is usually best to stick with your first guess.

Review.

The exam allows candidates to review and change answers within the time limit. Utilize this feature to mark troubling questions for review upon completing the rest of the exam.

Reference material.

The General Structures and the Mechanical & Electrical Systems divisions include reference materials accessible through an on-screen icon. These materials include formulas and other reference content that may prove helpful when answering questions in these divisions. Note that candidates may *not* bring reference material with them to the testing center.

Calculator.

Candidates must bring their own calculator to the testing center. Note that only nonprogrammable, noncommunicating, nonprinting calculators are allowed. Candidates will need only a basic scientific calculator with trigonometry functions.

Best answer questions.

Many candidates fall victim to questions seeking the "best" answer. In these cases, it may appear at first glance as though several choices are correct. Remember the importance of reviewing the question carefully and interpreting the language literally. Consider the example below.

Which of these cities is located on the east coast of the United States?

A. Boston

B. Philadelphia

C. Washington, DC

D. Atlanta

At first glance, it may appear that all of the cities could be correct answers. However, if you interpret the question literally, you'll identify the critical phrase as "on the east coast." Although each of the cities listed is arguably an "eastern" city, only Boston sits on the Atlantic coast. All the other choices are located in the eastern part of the country, but are not coastal cities.

ACKNOWLEDGMENTS

This introduction was written by John F. Hardt, AIA. Mr. Hardt is a principal of the firm Andrews Architects, Inc. in Columbus, Ohio. He is a graduate of Ohio State University (MArch) and has been in practice for more than 12 years.

ABOUT KAPLAN

Thank you for choosing Kaplan AEC Education as your source for ARE preparation materials. Whether helping future professors prepare for the GRE or providing tomorrow's doctors the tools they need to pass the MCAT, Kaplan possesses more than 50 years of experience as a global leader in exam prep and educational publishing. It is that experience and history that Kaplan brings to the world of architectural education, pairing unparalleled resources with acknowledged experts in ARE content areas to bring you the very best in licensure study materials.

Only Kaplan AEC offers a complete catalog of individual products and integrated learning systems to help you pass all nine divisions of the ARE. Kaplan's ARE materials include study guides, mock exams, question-and-answer handbooks, video workshops, and flash cards. Products may be purchased individually or in division-specific learning systems to suit your needs. These systems are designed to help you better focus on essential information for each division, provide flexibility in how you study, and save you money.

To order, please visit *www.KaplanAEC.com* or call (800) 420-1429.

The following symbols and abbreviations are used in this book and are generally understood in structural design practice.

Symbol or Abbreviation	**Meaning**
ft. or '	foot
ft^2 or sq. ft.	square foot
ft^3 or cu. ft.	cubic foot
ft.-kip or ft.-k or 'k	foot-kip
ft.-lb. or ft-# or '#	foot-pound
in. or "	inch
in^2 or sq. in.	square inch
in^3 or cu. in.	cubic inch
in.-kip. or in.-k or "k	inch-kip
in.-lb. or in-# or "#	inch-pound
kip or k	kip (1 kip = 1 kilo pound or 1000 pounds)
ksi or k/in^2	kips per square inch
lb. or #	pound
lb./cu. ft. or $\#/ft^3$ or pcf	pounds per cubic foot
plf or #/' or #/ft.	pounds per lineal foot
psf or $\#/ft^2$	pounds per square foot
psi or $\#/in^2$	pounds per square inch
Δ (delta)	1. total strain (deformation)
	2. thermal expansion or contraction
	3. deflection
θ (theta)	a common designation for an angle
π (pi)	the ratio of the circumference of a circle to its diameter, equal to 3.14159
Σ (sigma)	summation of
ϕ (phi)	strength reduction factor in reinforced concrete design
#	pounds

1. Two framing plans are shown below. Compared to the girders in plan A, how much greater is the required section modulus of the girders in plan B?

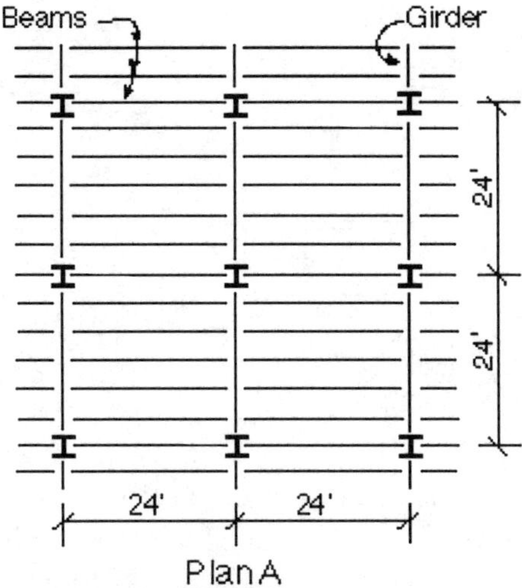

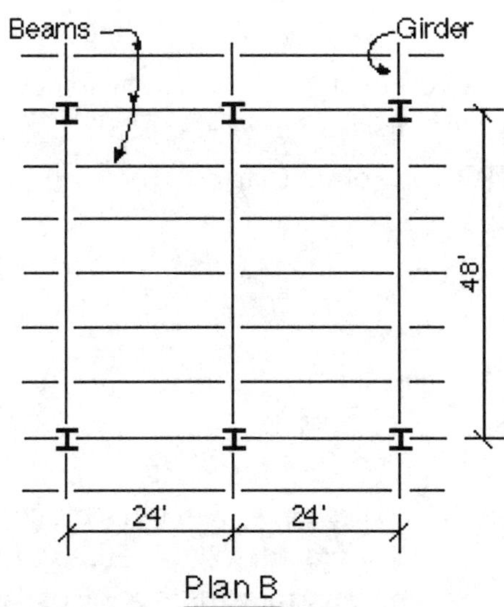

A. 50 percent greater

B. Twice as great

C. Four times as great

D. Eight times as great

2. All of the following are types of stress, *EXCEPT*

A. shear.

B. tension.

C. compression.

D. strain.

3. The load capacity of a structural steel column depends on the slenderness ratio KVr. In this ratio, K depends on

A. the length of the column.

B. the grade of steel.

C. the moment of inertia and area of the column.

D. the end conditions of the column.

4. A wide flange floor beam in a building is required to support a new piece of equipment, which will overstress the beam in bending. It is therefore necessary to strengthen the beam. Access is only from below. Which of the methods shown would be most effective, assuming there is sufficient headroom?

A.  Weld a horizontal plate to the bottom flange.

B. 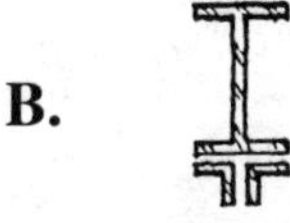 Weld two angles to the bottom flange.

C. 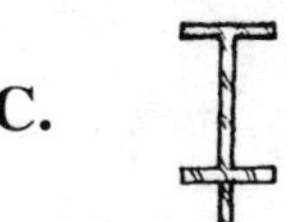 Weld a vertical plate to the bottom flange.

D. 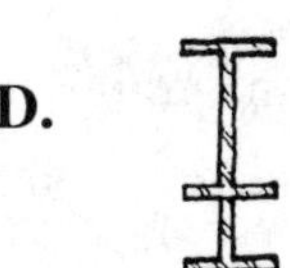 Weld a T plate to the bottom flange.

5. Which of the following is the correct moment diagram for the beam shown?

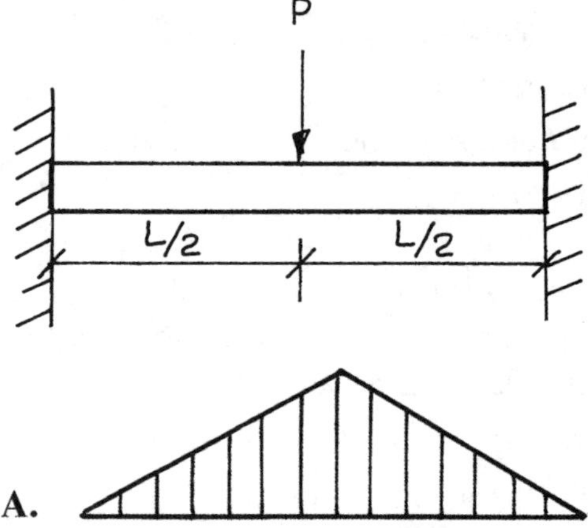

A. 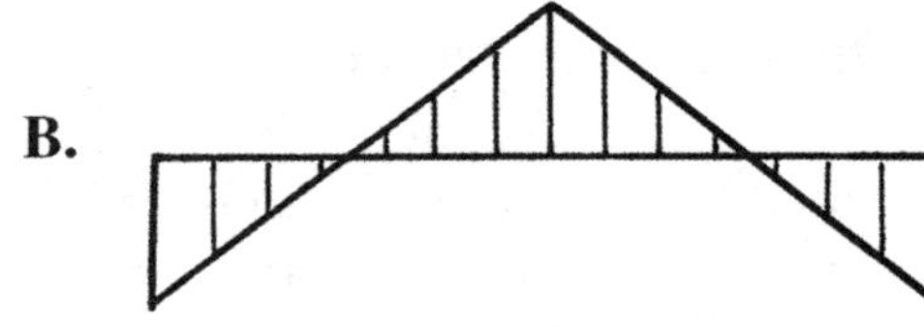

B.

C. 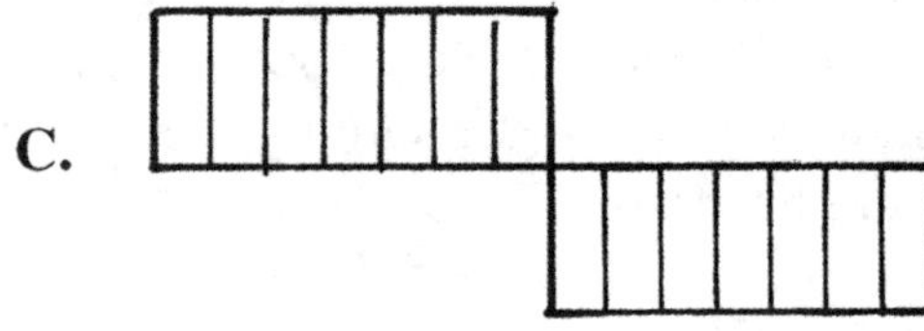

D. 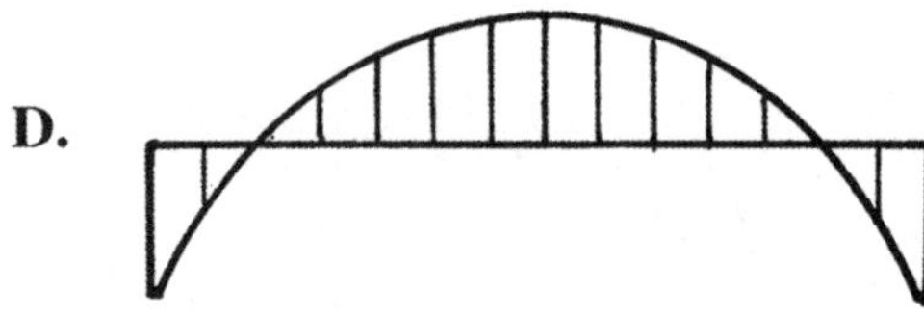

6. A square column pad is overstressed in shear. A solution is to

A. increase the pad size.

B. increase the area of reinforcing steel.

C. use more bars of smaller diameter.

D. increase the pad thickness.

7. Which of the diagrams below best represents the distribution of flexural stress in a homogeneous rectangular beam?

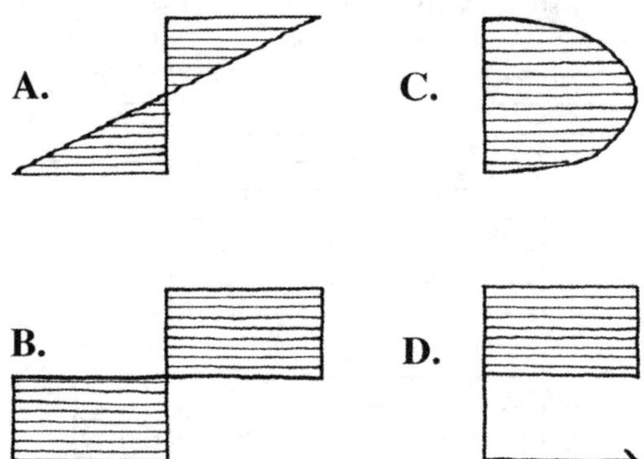

8. Truss members are generally subject to internal stresses that are axial. Which of the following will also cause bending stresses in truss members?

I.   Using trusses without diagonals

II. Closely spaced joists that span between trusses

III. Truss joints that provide restraint against rotation

A. I only

B. I and II

C. II and III

D. I, II, and III

9. During the design of a building, the deflection of a beam is calculated to be 0.90". In order to limit the maximum deflection of the beam to ¾ inch, how should the design be changed?

A. Substitute a beam having a section modulus 20 percent greater.

B. Substitute a beam having a moment of inertia 20 percent greater.

C. Substitute a beam having a moment of inertia 83 percent greater.

D. Substitute a beam having a yield point 20 percent greater.

10. Given the following reinforcing bar data:

Bar Size	Area
#4	0.20 in.^{2}
#5	0.31 in.^{2}
#6	0.44 in.^{2}
#7	0.60 in.^{2}
#8	0.79 in.^{2}
#9	1.00 in.^{2}

If the ultimate tensile capacity of 2-#6 grade 40 reinforcing bars is compared with that of 1-#8 grade 60 reinforcing bar,

A. the #6 bars have the greater capacity.

B. the #8 bar has the greater capacity.

C. the capacities of both sets of bars are identical.

D. the capacity cannot be determined without knowing the 28-day compressive strength (f$'_c$) of the concrete in which the reinforcing bars are used.

11. Which of the following statements concerning the flat plate floor system are generally correct?

I.   The construction depth is shallow.

II. Shear stresses near the columns are high.

III. The system is inherently very stiff.

IV. Beams between the columns are required in both directions.

V. The system is economical for live loads of 100 psf or more and for spans of about 30 feet.

A. I and II only

B. II, III, and V

C. I, IV, and V

D. I, II, and V

12. Considering the two beams shown below, the three-span continuous beam and the simple beam, which of the following is *NOT* a correct statement?

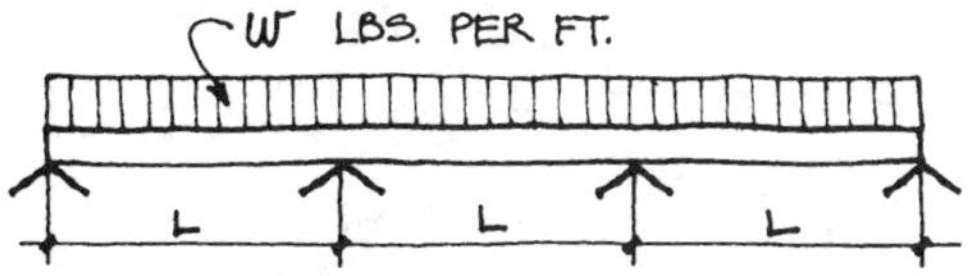

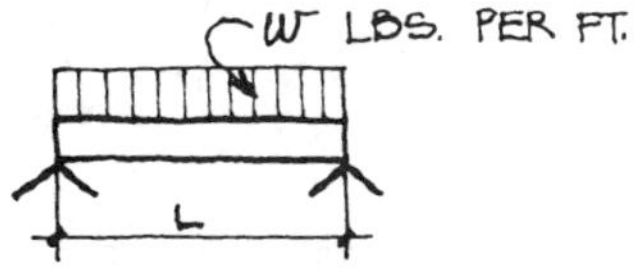

A. The maximum positive bending moment in the simple beam is greater than in the continuous beam.

B. The maximum negative bending moment in the continuous beam is greater than in the simple beam.

C. The maximum positive bending moment is greater in the end spans of the continuous beam than in the center span.

D. The maximum deflection of the continuous beam is the same as that of the simple beam.

13. In a simple beam supporting a concentrated load, as shown below, where do the beam fibers lengthen and where do they shorten?

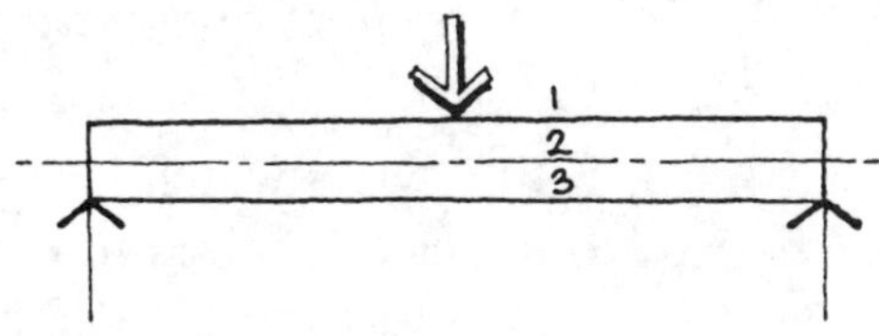

A. Lengthen at 1, shorten at 3

B. Lengthen at 3, shorten at 1

C. Lengthen at 1 and 2, shorten at 3

D. Lengthen at 2 and 3, shorten at 1

14. What is the total stress in member *a* of the cantilever truss shown below?

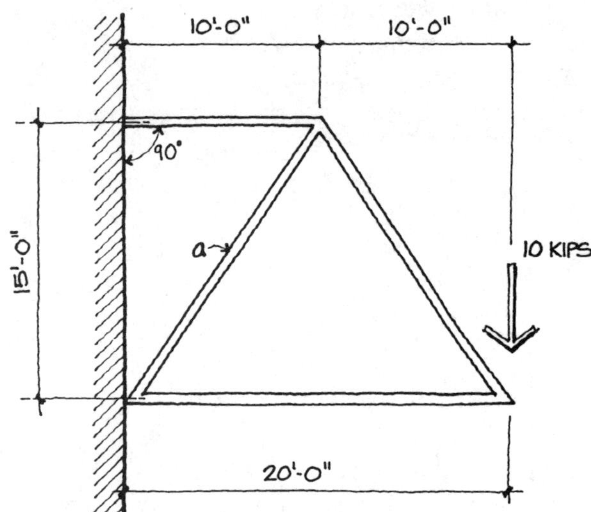

A. 12.02 kips tension

B. 12.02 kips compression

C. 6.67 kips compression

D. 13.33 kips compression

15. The weld symbol shown indicates

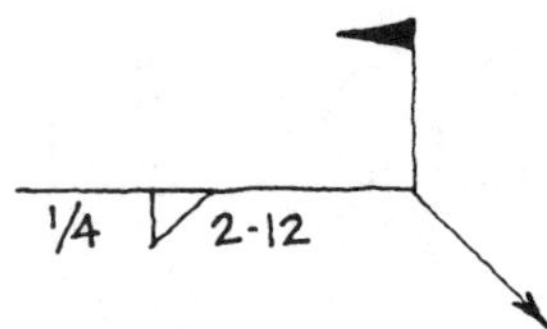

A. a 1/4 inch intermittent bevel weld to be made in the field.

B. a 1/4 inch intermittent fillet weld to be made in the field.

C. a 1/4 inch fillet weld which has been flagged to indicate required ultrasonic testing of 2 out of every 12 welds.

D. a 1/4 inch fillet weld to be made in the field with class 2-12 electrodes.

16. In general, the most economical reinforced concrete framing system is one in which

A. the amount of reinforcing steel is kept to a minimum by increasing the sizes of the members.

B. the vertical loads are resisted by bearing walls and the lateral wind or earthquake loads by rigid frames.

C. the areas in which the concrete is in tension are replaced with voids, as in a waffle slab, thus reducing the weight of the system.

D. the forming is uniform and repetitive and the reinforcing layout is simple and repetitive.

17. What is the left reaction for the beam shown below?

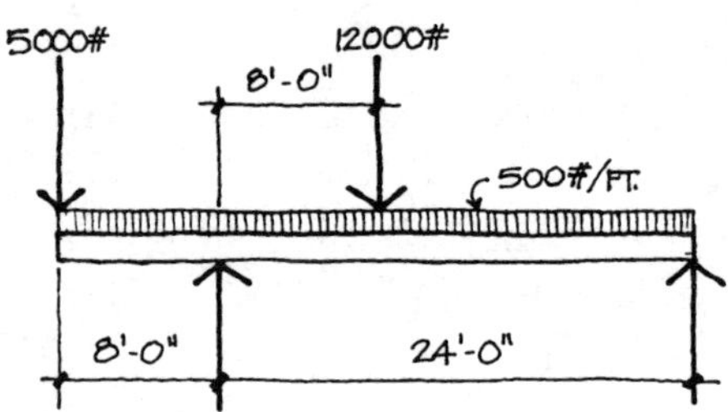

A. 25,333#

B. 7,667#

C. 21,333#

D. 23,000#

18. Each of the following items is associated with Roman architecture, *EXCEPT*

A. thermae.

B. basilica.

C. amphitheater.

D. stoa.

19. A beam is loaded as shown. Which of the following is the corresponding shear diagram?

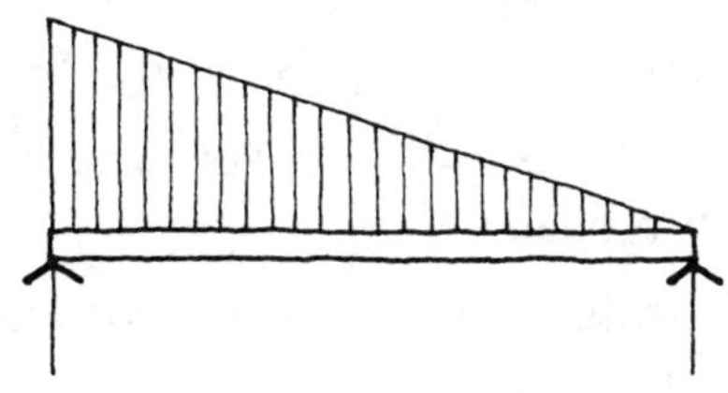

A.

B.

C.

D.

20. Which of the following statements concerning the dome of the Florence Cathedral is *NOT* correct?

A. It was designed by the great Renaissance architect Brunelleschi.

B. It consists of a thick inner dome connected to a thinner outer dome.

C. The construction of the dome required the largest temporary structure ever used up to that time for shoring.

D. A series of built-in iron chains act to resist tensile forces and prevent the dome from cracking or collapsing.

21. Select the correct statement about the Pantheon.

A. Built-in iron chains were used to resist hoop tensile stresses in the dome.

B. Its dome was constructed of individual stones carefully fitted together.

C. It was destroyed in the great earthquake of 1847.

D. Its concrete dome, the largest of antiquity, still stands today.

22. Which of the following men were important figures in the evolution of the skyscraper?

I.   Elisha Graves Otis

II. Fazlur Khan

III. William LeBaron Jenney

IV. Robert Maillart

A. I and III

B. II and IV

C. I, II, III, and IV

D. I, II, and III

23. The roof of the Dulles Airport terminal is supported by a series of monumental concrete piers. Why do these piers lean outward?

A. To express the concept of flight

B. To counteract the pull of the roof cables

C. To counteract the thrust of the roof arches

D. To provide greater resistance to lateral forces from wind or earthquake

24. When the outside temperature increases, the top of a dome will

A. move up.

B. move down.

C. remain in the same location.

D. move sideways.

25. The John Hancock Building in Chicago and the Sears Tower have what in common?

    I.   They were designed by Skidmore, Owings, and Merrill.

    II. They employ tubular systems to resist lateral loads.

    III. They utilize reinforced concrete framing.

    A. I only

    B. II only

    C. I and II

    D. II and III

26. Which of the following statements about live load is correct?

    A. In order to provide an adequate factor of safety, the live load may not be less than the dead load.

    B. Live load is the load superimposed by the use and occupancy of the building, not including the wind load, earthquake load, or dead load.

    C. If approved by the building official, the live load may be neglected in the design of a building, if it is less than the dead load.

    D. Because of their greater vulnerability to failure, long-span structures must be designed for a greater live load than conventional structures.

27. Thin shells and membranes are able to resist load because of their

    A. compressive strength.

    B. shear strength.

    C. form.

    D. tensile strength.

28. Select the correct statements about the horizontal thrust at the base of a three-hinged arch.

    I.   The thrust is directly proportional to the vertical load supported by the arch.

    II. The thrust is directly proportional to the arch span.

    III. The thrust is directly proportional to the rise of the arch.

    IV. The thrust increases if the temperature rises and decreases if the temperature drops.

    A. I, II, and IV

    B. I and III

    C. II and IV

    D. I and II

29. Select the *INCORRECT* statement about long-span construction.

    A. Secondary stresses in long-span structures may justify an increase in factors of safety.

    B. Consideration should be given to using different factors of safety for site-assembled and factory-assembled building components.

    C. The factors of safety for long-span building components should not exceed those specified in the building code.

    D. The effects of temperature are more critical in long-span structures than in conventional structures.

30. The structural action of a membrane is based on its

    A. tensile strength.

    B. shear strength.

    C. bending strength.

    D. compressive strength.

31. The U.S. Pavilion at Expo 70 in Osaka had an air-supported roof spanning 262 feet by 460 feet. Its occupied space was pressurized, and its roof was subject to upward suction from wind. Its concrete foundation ring was primarily subject to what type of stress?

    A. Compression

    B. Tension

    C. Shear

    D. Flexure

32. A series of parallel arches, skewed with respect to the axes of a building, which intersect another series of skewed arches, is known as a

    A. groined vault.

    B. lamella.

    C. Schwedler dome.

    D. hyperbolic paraboloid.

33. Which of the following statements are true? A long-span shell structure

    I. is suitable for resisting concentrated loads.

    II. can resist both compression and tension in its own plane.

    III. cannot resist any substantial bending stress.

    IV. cannot resist any shear stress.

    A. II, III, and IV

    B. I and II

    C. III and IV

    D. II and III

34. The four basic structural systems shown below are being considered for a long-span roof. Which of them supports uniform vertical loads primarily by compression?

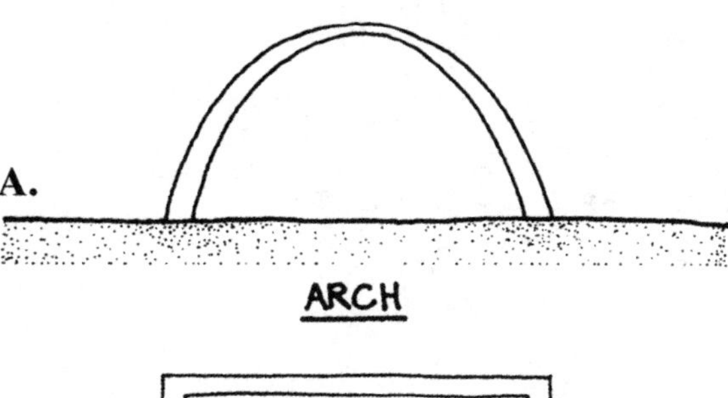

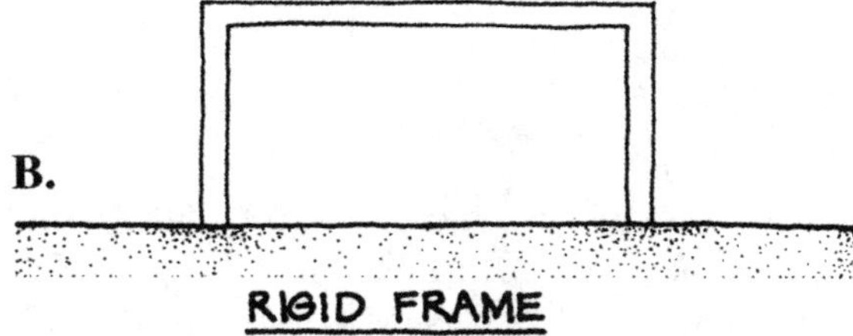

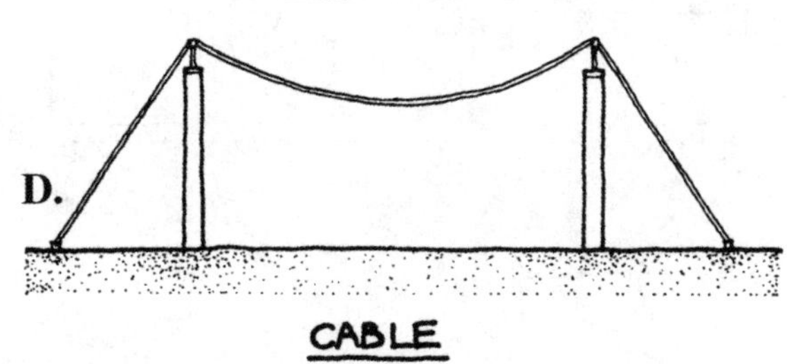

35. Which of the following is the best definition of "space frame"?

    A. A two-way system of rigid frames.

    B. A moment-resisting frame.

    C. A two-way system of trusses.

    D. Any three-dimensional structural system.

36. Radial cables support the roof of a circular building, as shown below. Which of the following statements is correct concerning the stress in the inner and outer rings?

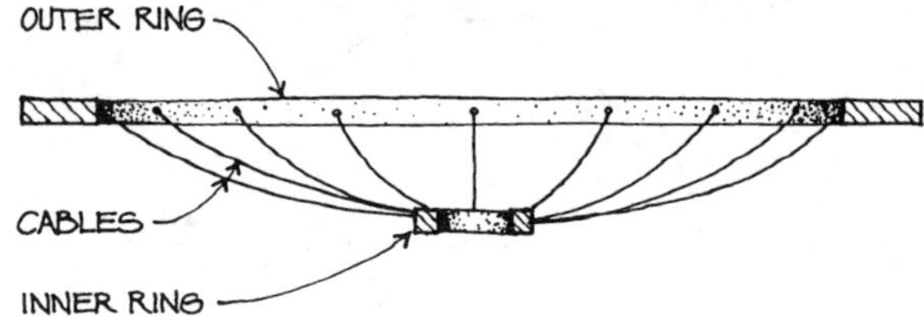

    A. Both inner and outer rings are in tension.

    B. Both inner and outer rings are in compression.

    C. The inner ring is in tension and the outer ring is in compression.

    D. The inner ring is in compression and the outer ring is in tension.

37. Select the correct statement concerning the structural member shown.

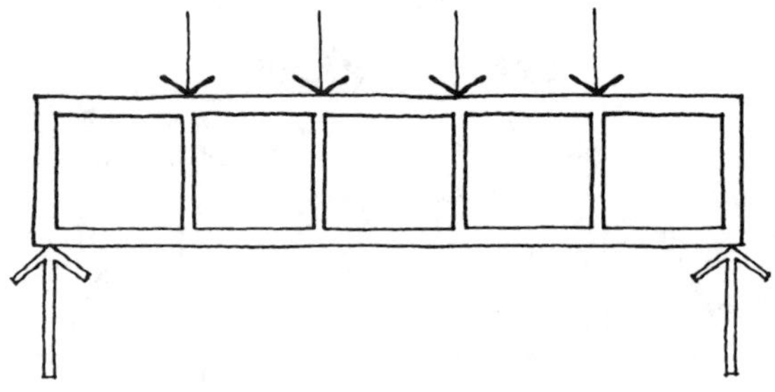

    A. It tends to deflect less than a comparable triangulated truss.

    B. It is unstable, as it is not composed of triangles.

    C. In addition to axial stresses, the chord members must also resist bending moments, which are greater close to the supports than at midspan.

    D. The joints must be pinned, to avoid excessive stresses caused by deformation.

38. For which of the following building types is long-span construction likely to be used?

    I.   Exhibition hall

    II. Basketball pavilion

    III. Library

    IV. Auditorium

    V. University science building

    VI. Municipal administration building

    A. I, II, and IV

    B. II, III, and V

    C. I, IV, and VI

    D. I, II, III, IV, and V

39. Which surface in reinforced concrete can be formed using straight planks?

    A. Hyperbolic paraboloid

    B. Folded plate

    C. Geodesic dome

    D. Lamella

40. In what system used in Gothic cathedrals is load supported by piers at its four corners?

    A. Lamella

    B. Groined vault

    C. Dome

    D. Folded plate

41. What is the essential structural element of all suspension structures?

    A. Arch

    B. Dome

    C. Cable

    D. Plate girder

42. What system can resist both horizontal and vertical loads because the angle between its horizontal and vertical elements cannot change?

   A. Vierendeel truss

   B. Tapered girder

   C. Space frame

   D. Rigid frame

43. A rigid frame with hinged bases supports a uniformly distributed vertical load as shown above. Which of the following correctly shows the shape of the deflected frame?

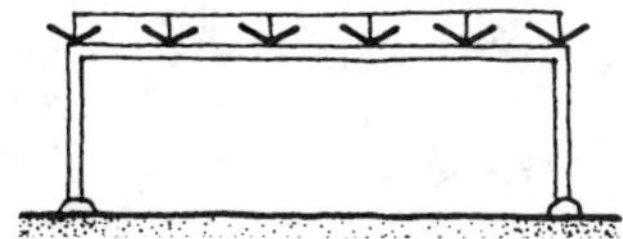

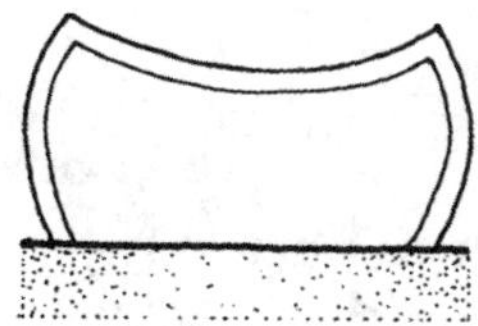
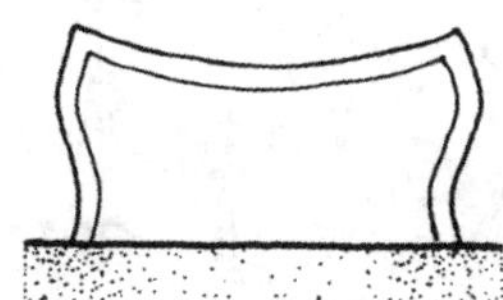

A.   **C.**

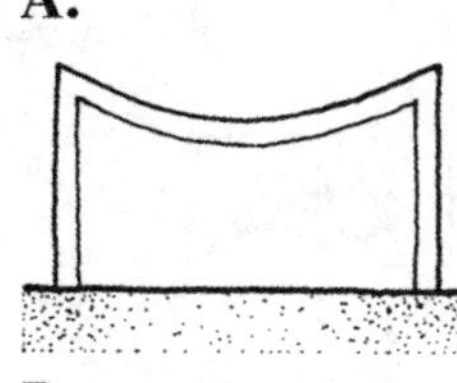
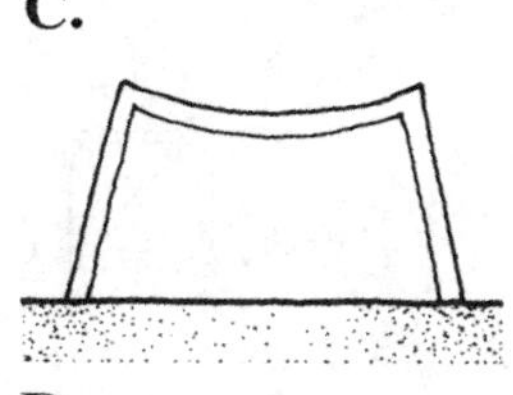

B.   **D.**

44. What system can have almost any configuration, so long as it is composed of triangles?

   A. Lamella

   B. Geodesic dome

   C. Truss

   D. Space frame

45. Select the correct statements about the three-hinged arch shown.

   I. It is statically determinate.

   II. It is capable of supporting both vertical and horizontal loads.

   III. The connection at the center must be able to resist moment.

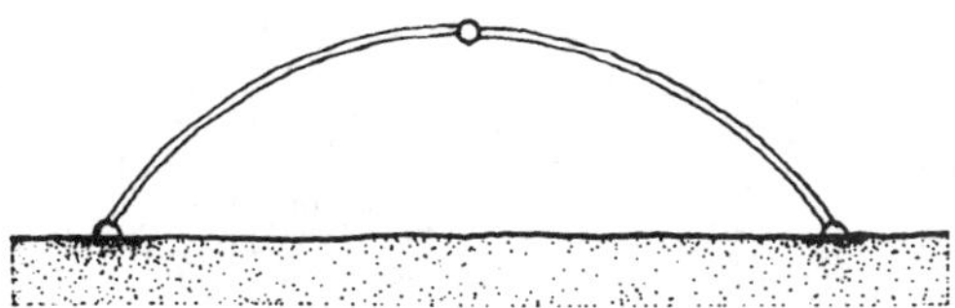

   A. I only

   B. II only

   C. I and II

   D. I, II, and III

46. When a member undergoes stress, its length changes. If the member does not return to its original length when the stress is removed, the action is called

   A. elastic.

   B. ductile.

   C. inelastic.

   D. malleable.

47. A 15-foot-long pin-ended column of A36 steel supports a load of 200 kips. The column's dimensions and properties are as follows: area = 14.1 in.^{2}, depth = 8.50", web thickness = 0.400", flange width = 8.110", flange thickness = 0.685", $r_{x\text{-}x}$ = 3.6", $r_{y\text{-}y}$ = 2.08". What is the compressive stress in the column?

   A. 2.9 ksi

   B. 14.2 ksi

   C. 22.3 ksi

   D. 24.0 ksi

48. Four different girder arrangements are being considered to support three equal 60-foot spans, as shown below. Which arrangement will result in the *LEAST* deflection and moment in span 1-2?

A.

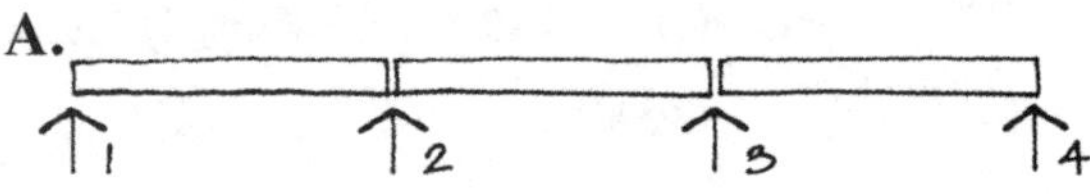

B.

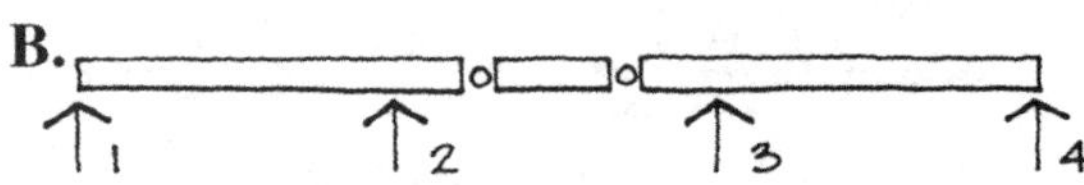

C.

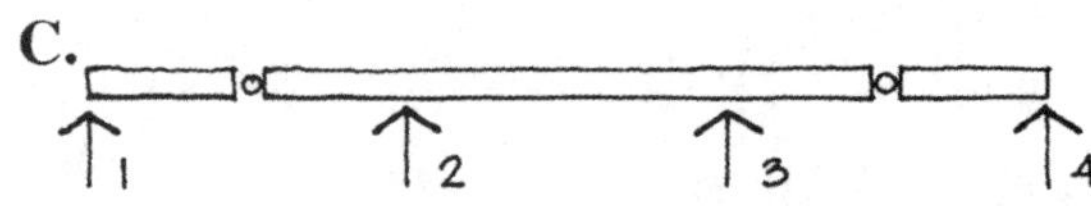

D.

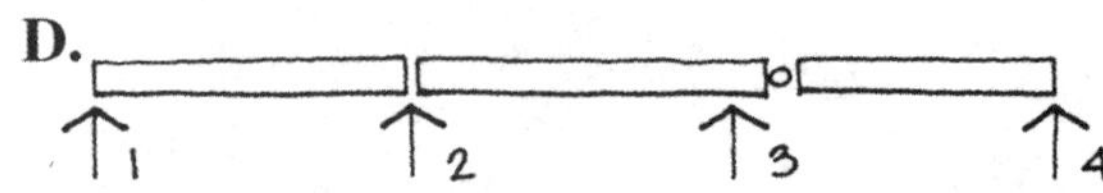

49. Loads applied to a truss chord between panel points result in

   A. local buckling of the chord.

   B. combined bending and axial stress in the chord.

   C. high shear stresses in the web members.

   D. compression in the chord.

50. Hooke's Law states that stress is proportional to strain. The constant ratio of stress to strain is termed

   A. the modulus of rupture.

   B. the modulus of elasticity.

   C. stiffness.

   D. the coefficient of expansion.

51. A truss is loaded as shown. Which sketch below correctly shows the reactions at A and B?

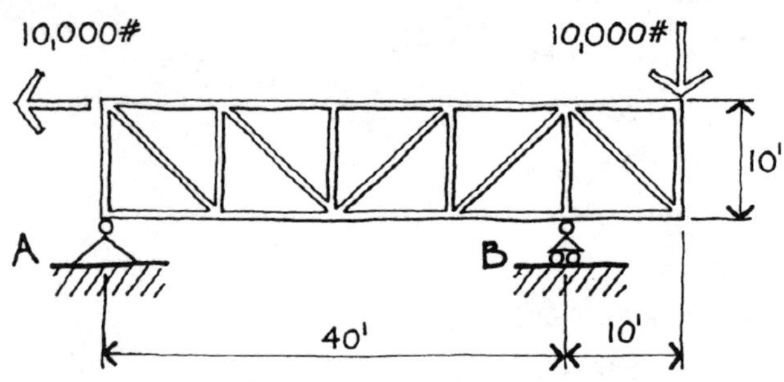

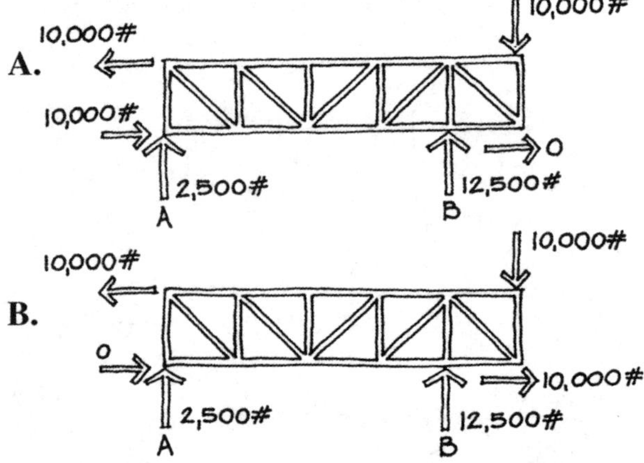

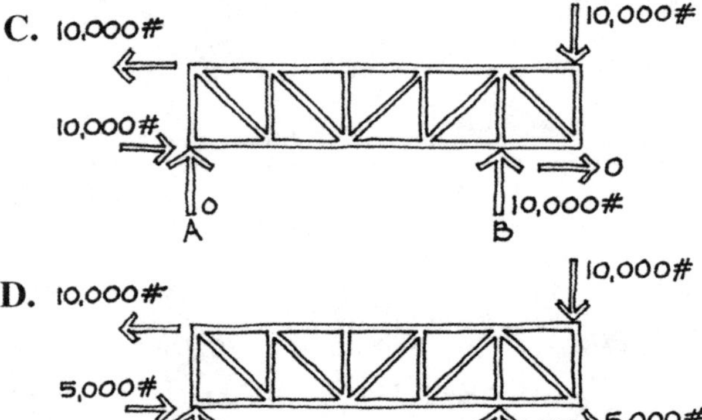

52. A one-story rigid frame with hinged bases supports a uniformly distributed load as shown below. Which of the following is the correct moment diagram for the frame?

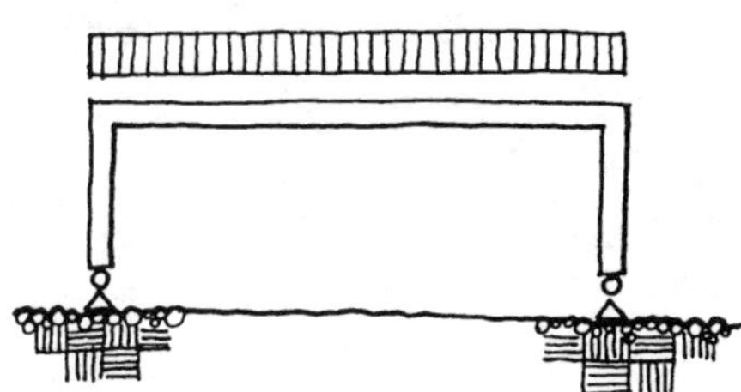

A.

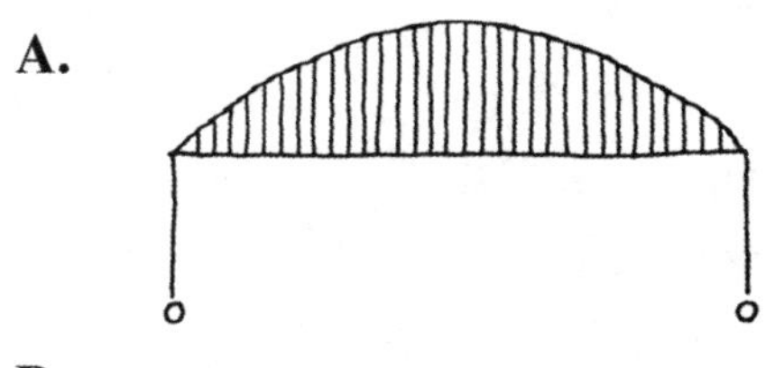

B.

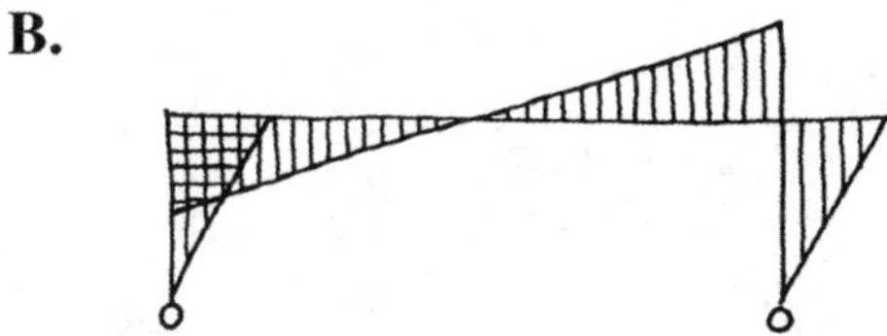

C.

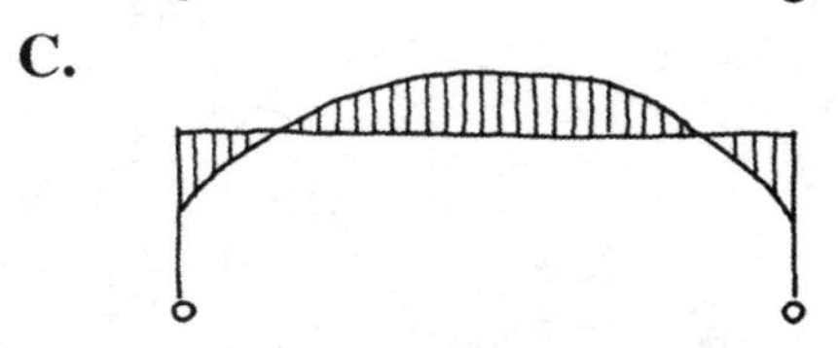

D.

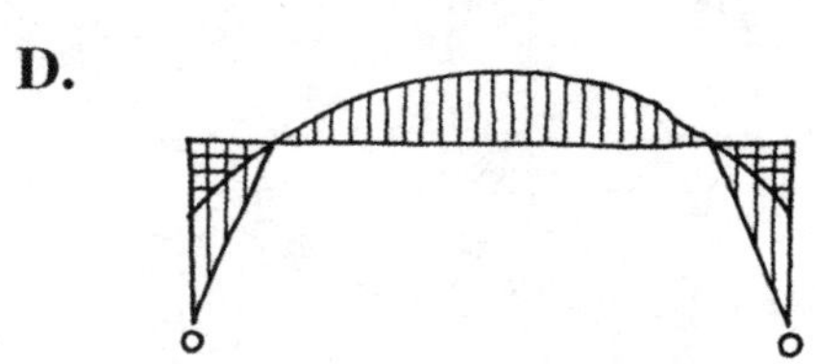

53. The ability of a structural material or system to sustain inelastic deformations without collapse is called

A. ductility.

B. strength.

C. yielding.

D. stiffness.

54. The two-span continuous beam shown supports a uniform load on span AB only and is connected to supports A, B, and C. Neglecting the weight of the beam, the reaction at C

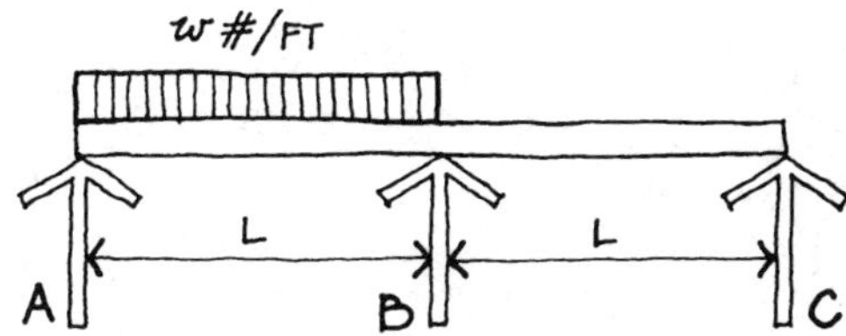

A. acts upward.

B. acts downward.

C. may act upward or downward, depending on the magnitude of the uniform load.

D. is zero.

55. Which of the following diagrams best shows the deflected shape of the rigid frame shown below?

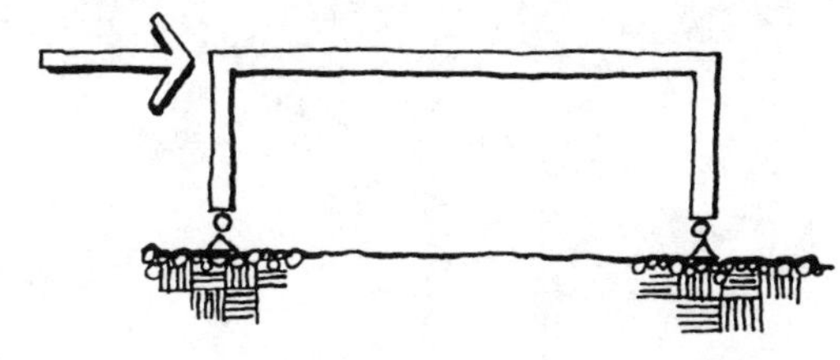

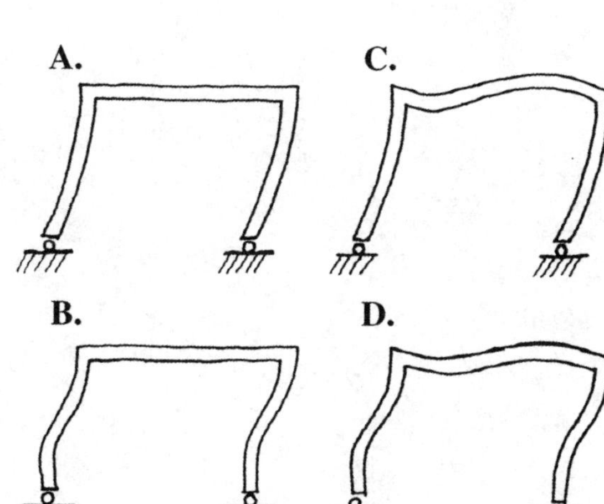

A.

C.

B.

D.

56. A rigid bar 20 feet long supports a load of 1,000 pounds at its end as shown. Which of the following sketches correctly represents the horizontal and vertical components of the forces acting on the bar?

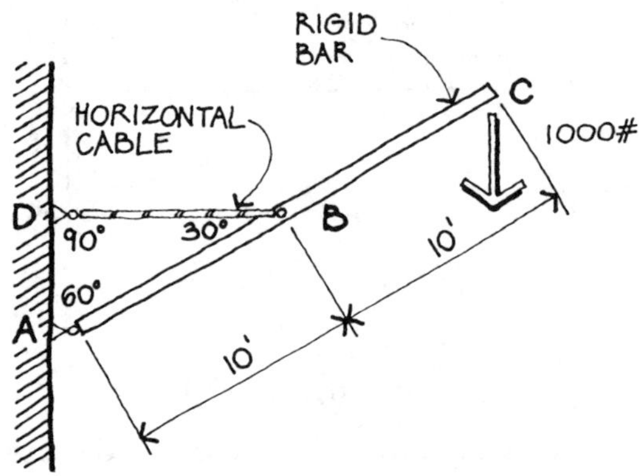

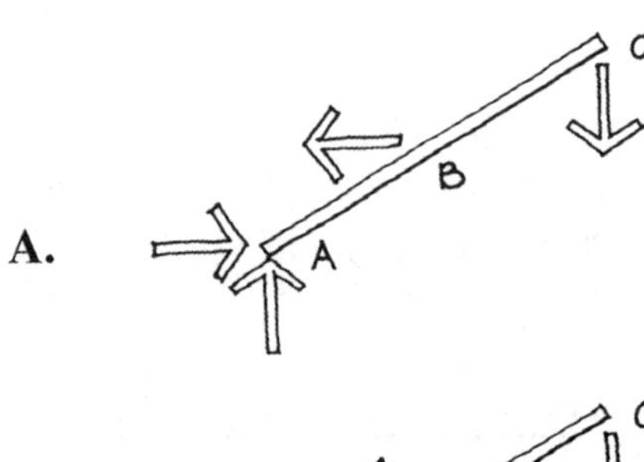

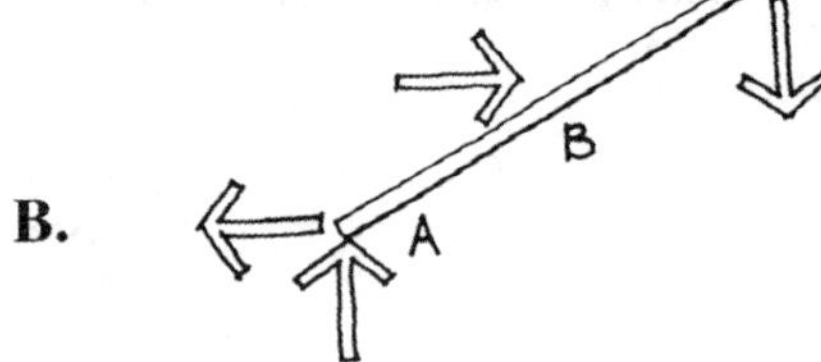

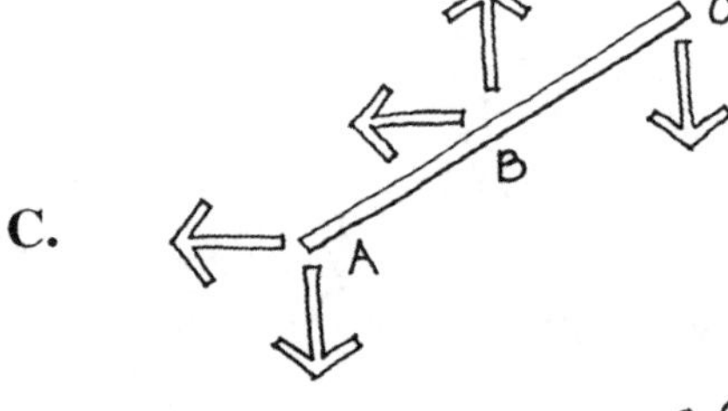

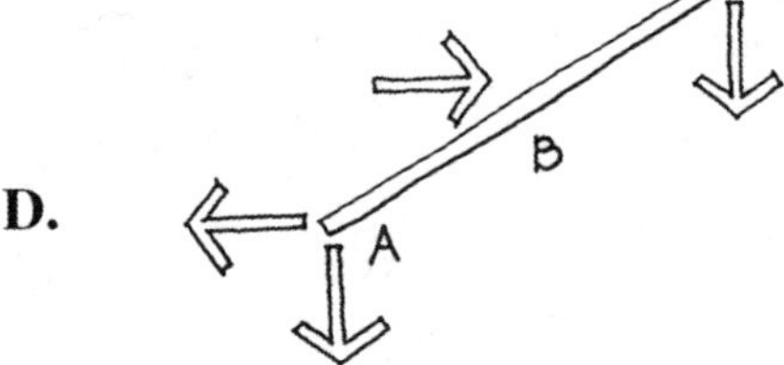

57. What is the internal force in the vertical member of the king post truss shown?

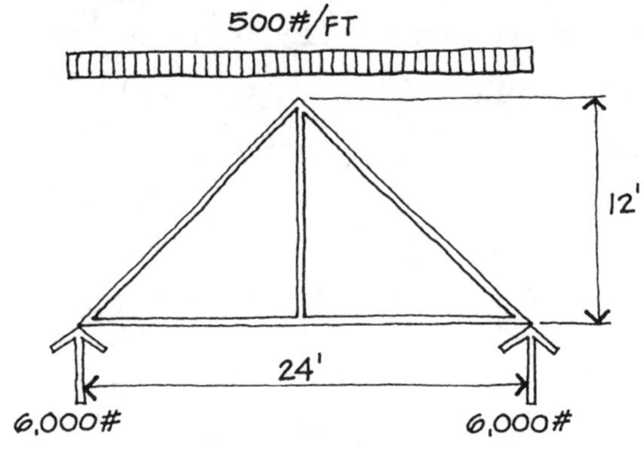

A. Zero

B. 3,000#

C. 6,000#

D. 12,000#

58. In the detail shown below, the wood members are loaded in

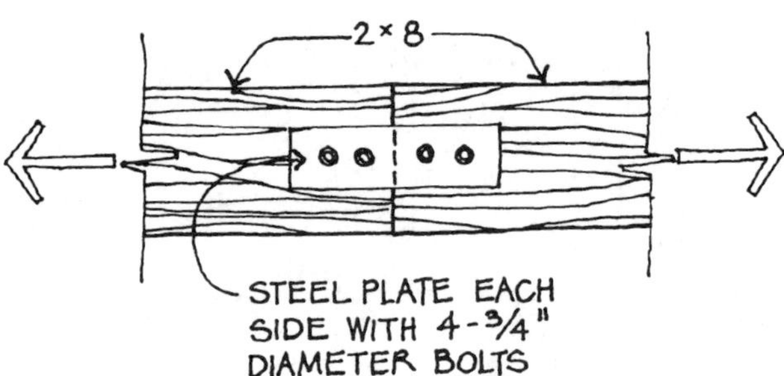

A. tension perpendicular to the grain.

B. tension parallel to the grain.

C. compression parallel to the grain.

D. horizontal shear.

59. A load of 50 pounds per lineal foot is applied to the top rail as shown. What is the force in the upper bolt?

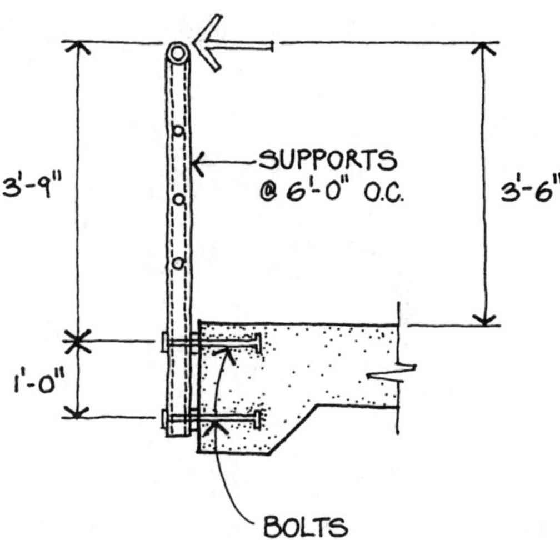

A. 238 lbs. tension

B. 1,125 lbs. tension

C. 1,425# tension

D. 1,425# compression

60. Select the *INCORRECT* statement about equilibrium.

A. If only two forces act on a body in equilibrium, they must be equal and opposite to each other.

B. If the vectorial sum of all the forces acting on a body in any direction equals zero, the body must be in equilibrium.

C. If three non-parallel forces acting on a body produce equilibrium, their lines of action must pass through a common point.

D. For equilibrium, the clockwise moments acting on a body must be equal to the counterclockwise moments.

61. A member in a testing machine elongates 0.03 inch when a tensile load of 75 kips is placed on it. The member is 12 inches long and one inch by one inch in cross-section. What is the modulus of elasticity of the material?

A. 2,500 ksi

B. 3,000 ksi

C. 25,000 ksi

D. 30,000 ksi

62. An opening must be provided in a reinforced concrete beam. Which letter in the sketch shows the location which will *LEAST* affect the beam's load-carrying capacity?

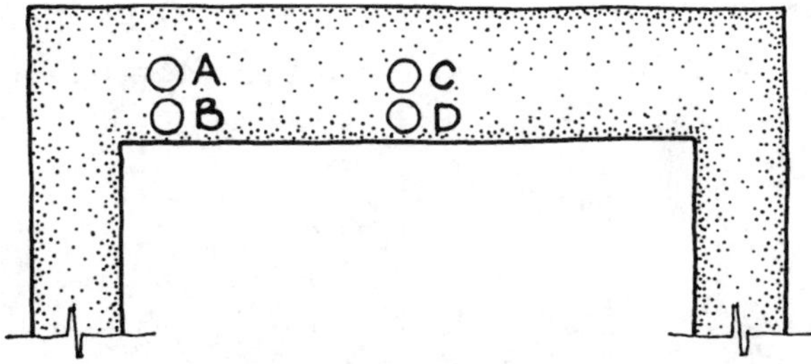

63. A post and beam system is able to resist

I. vertical live and dead loads.

II. lateral loads in its own plane.

III. lateral loads perpendicular to its own plane.

A. I only

B. I and II

C. II and III

D. I, II, and III

64. What is the maximum moment in the beam shown?

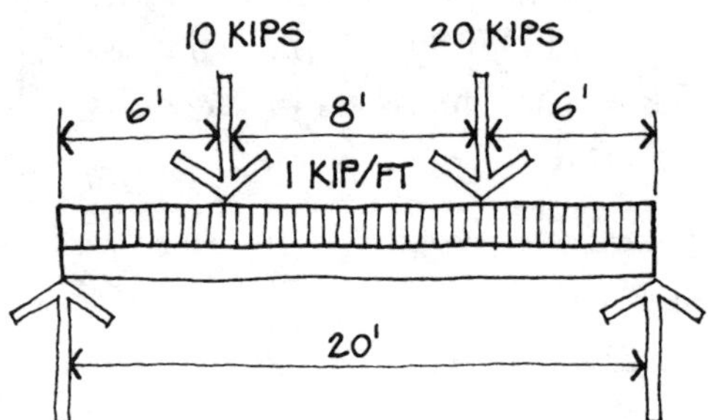

A. 50.0 ft.-kips

B. 110.0 ft.-kips

C. 144.5 ft.-kips

D. 170.0 ft.-kips

65. What is the moment of inertia of the section shown below about its x-x axis? $I = bd^3/12$.

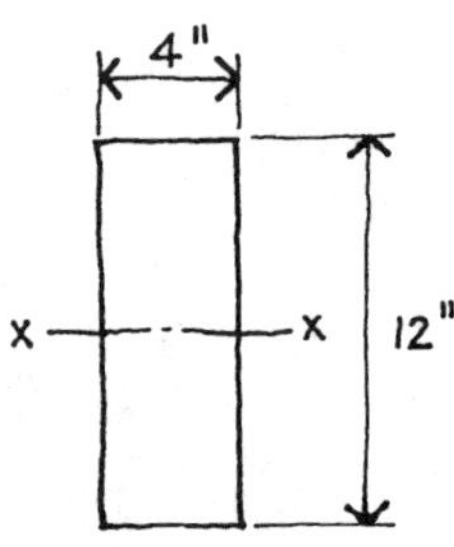

A. 48 in.^{4}

B. 64 in.^{4}

C. 96 in.^{4}

D. 576 in.^{4}

66. Building structures must be designed to resist

A. dead load plus half the live load.

B. live load plus half the dead load.

C. dead load plus live load.

D. dead load or live load, whichever is greater.

67. A load of 1,000 pounds is supported by cables A and B as shown. What is the tension in each cable?

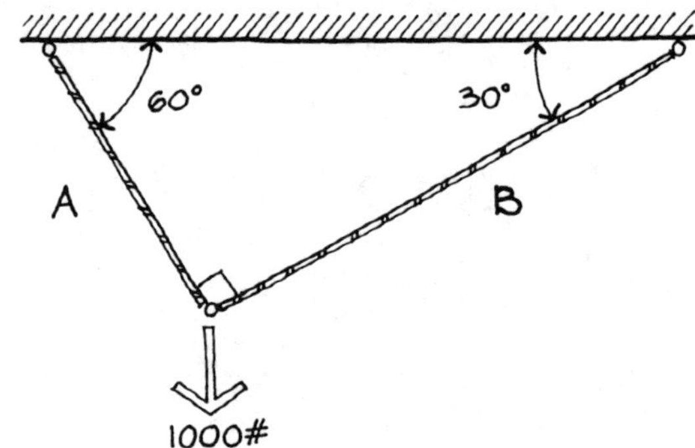

A. A = 500#, B = 500#

B. A = 500#, B = 866#

C. A = 866#, B = 500#

D. A = 866#, B = 866#

68. A 10-foot-deep tank is filled with water. The magnitude of the pressure exerted by the water against the sides of the tank depends on which of the following?

I.   The unit weight of the water

II. The depth of the water

III. The width of the tank

A. I and II

B. II and III

C. I and III

D. I, II, and III

69. Complete the following statement. An arch supports load by

A. compression.

B. tension.

C. bending.

D. a combination of compression and bending.

70. For the three-hinged arch shown, the horizontal thrust is equal to H. If the span length (L) is doubled, while the rise (h) and the load per lineal foot (w) remain the same, what will be the magnitude of the horizontal thrust?

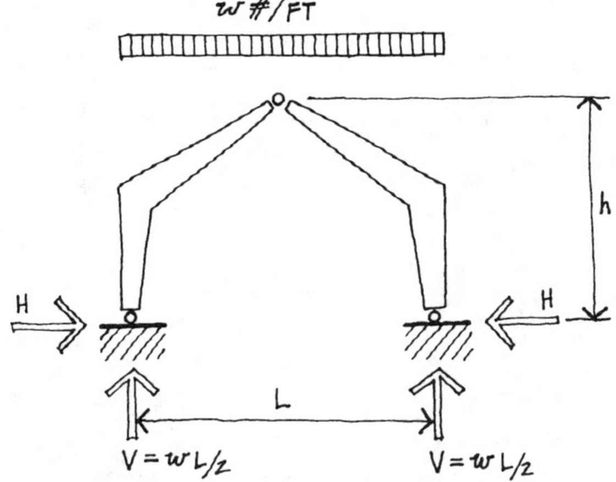

A. H

B. 2H

C. 3H

D. 4H

71. Which of the following diagrams correctly shows the variation in shear stress in a rectangular homogeneous beam?

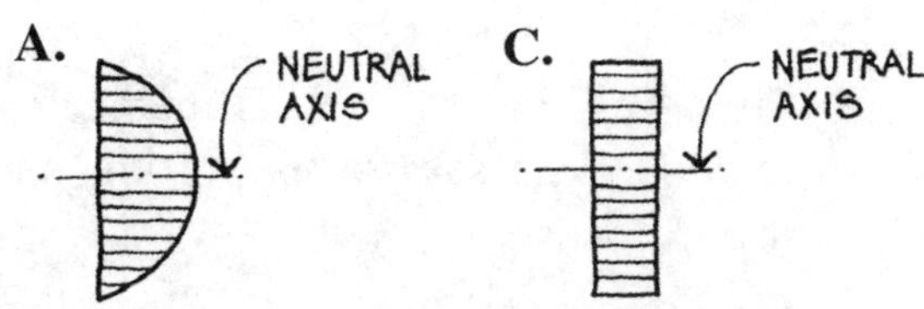

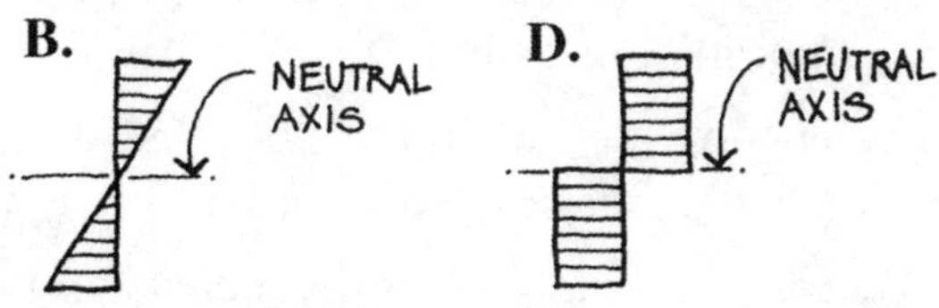

72. Wherever a hinge or pin exists in a structure, there can be no

A. shear.

B. moment.

C. tension.

D. compression.

73. What is the internal axial force in member *a* of the truss below?

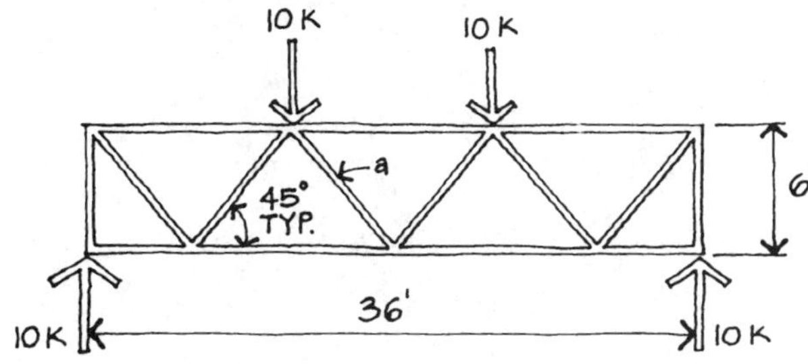

A. Zero

B. 5 kips compression

C. 7.07 kips compression

D. 10 kips compression

74. Complete the following statement. A cable supports load by

A. compression.

B. tension.

C. bending.

D. a combination of tension and bending.

75. For a simply supported glued laminated beam subject to vertical loads, where is the bending stress maximum?

A. At the outermost laminations near midspan

B. At the outermost laminations near the supports

C. At the neutral axis near midspan

D. At the neutral axis near the supports

76. A cantilever beam supports a uniformly distributed load as shown below. Which of the following is the appropriate moment diagram?

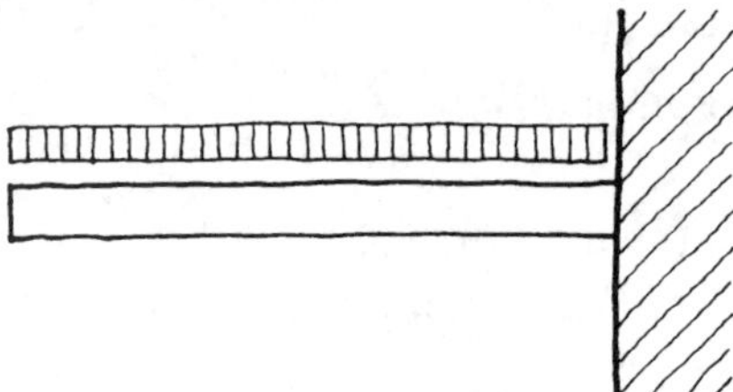

A.

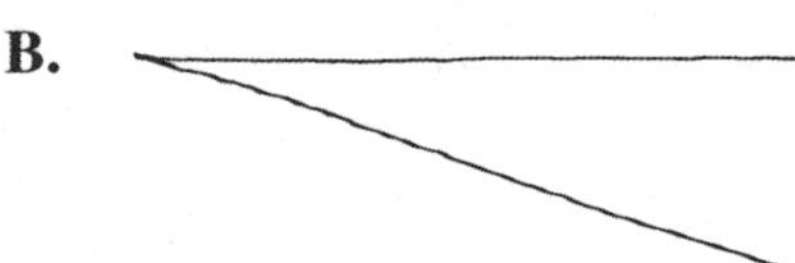

B.

C.

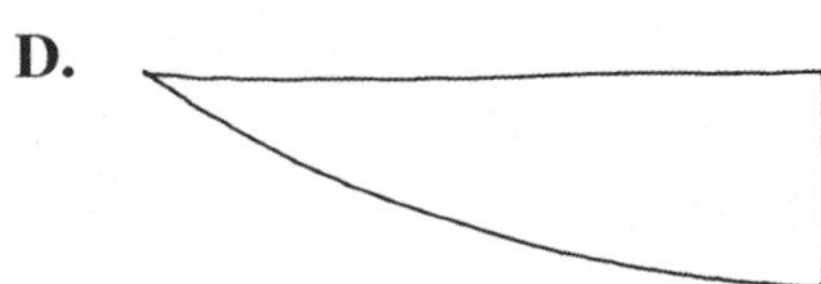

D.

77. The stress at which a ductile material continues to deform without an increase in load is called the

A. elastic limit.

B. yield point.

C. ultimate strength.

D. modulus of elasticity.

78. Which of the four beams below are statically indeterminate?

I.

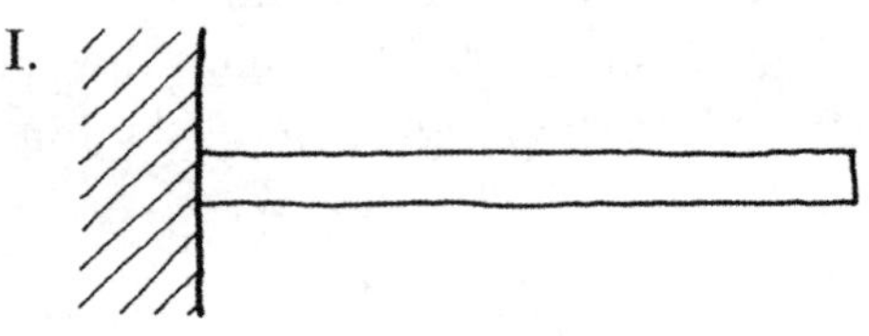

II.

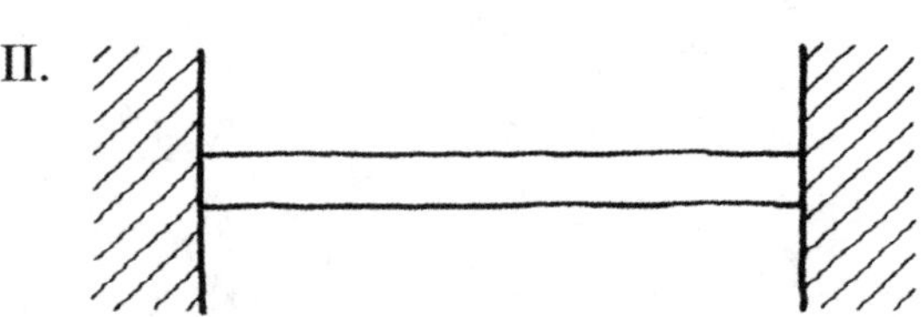

III.

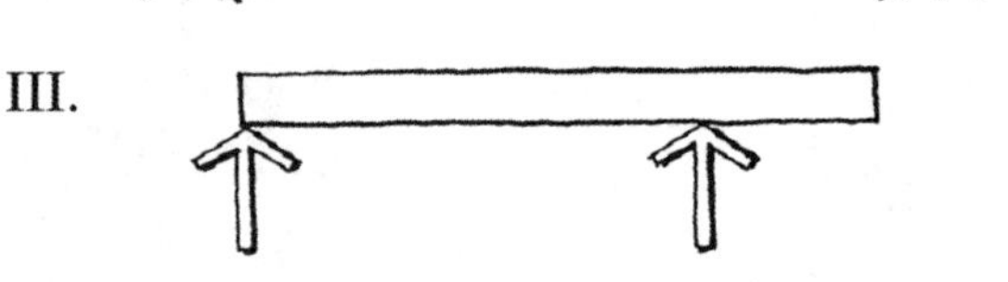

IV.

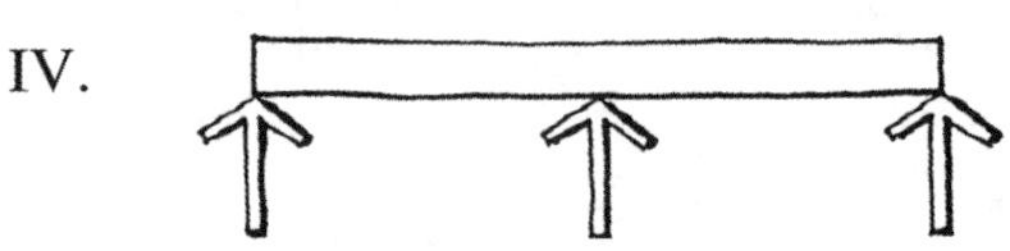

A. I, II, and IV

B. II and IV

C. III and IV

D. I, II, and III

79. What does redundancy in a structure refer to?

A. Providing more structural members than are necessary to support the anticipated loads

B. Having the ability to redistribute loads to other structural elements in case of overload or failure

C. Having a structural system that is statically indeterminate

D. Having main girders that consist of two members each

80. A beam supports two 10-kip concentrated loads as shown. What are the values of R_1 and R_2?

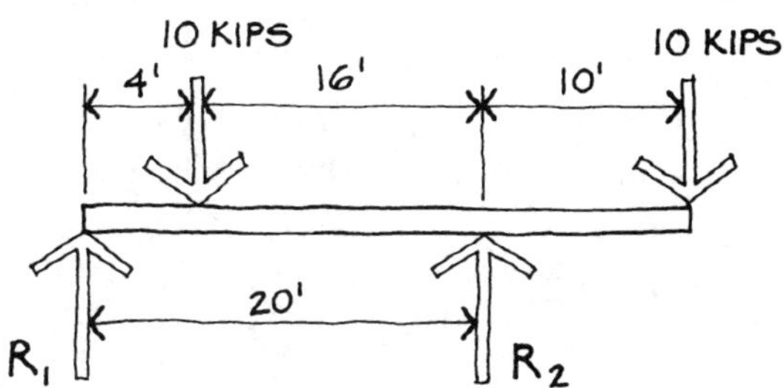

    A. $R_1 = 17$ kips, $R_2 = 3$ kips

    B. $R_1 = 3$ kips, $R_2 = 17$ kips

    C. $R_1 = 13$ kips, $R_2 = 7$ kips

    D. $R_1 = 7$ kips, $R_2 = 13$ kips

81. The section modulus of a beam is equal to its moment of inertia

    A. divided by the allowable flexural stress.

    B. multiplied by the allowable flexural stress.

    C. divided by the distance from the neutral axis to the outermost fiber.

    D. multiplied by the distance from the neutral axis to the outermost fiber.

82. Wide flange shapes are economical flexural members because

    A. the two flanges are equal in area.

    B. the flange width-to-thickness ratio is high.

    C. the web depth-to-thickness ratio is low.

    D. most of the beam's material is in the flanges.

83. What is the shear at section *x-x* of the beam shown?

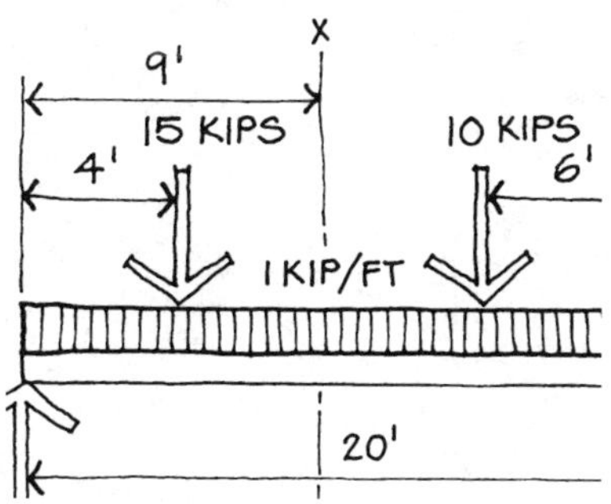

    A. Zero   **C.** 4,000#

    B. 1,000#   **D.** 5,000#

84. The shear diagram for a beam is shown. Which of the following is the corresponding moment diagram?

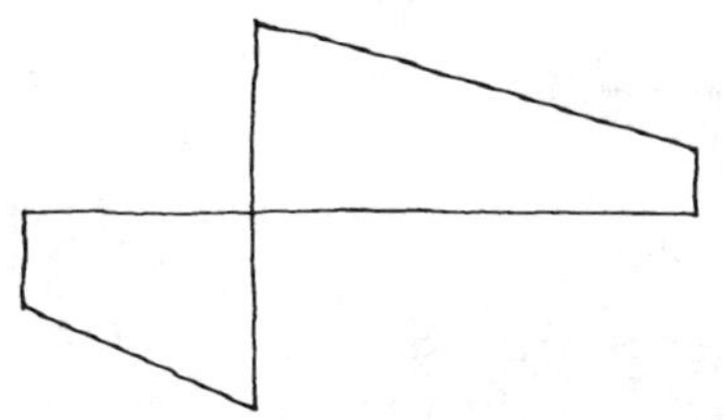

A. 

B. 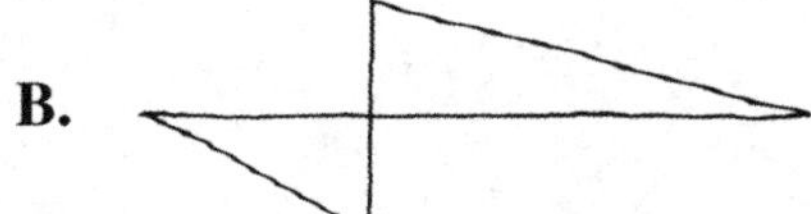

C. 

D. 

85. A one-story rigid frame with fixed bases resists a horizontal load as shown. Neglecting the weight of the frame, which of the following correctly shows the directions of the reactions and moments at the bases?

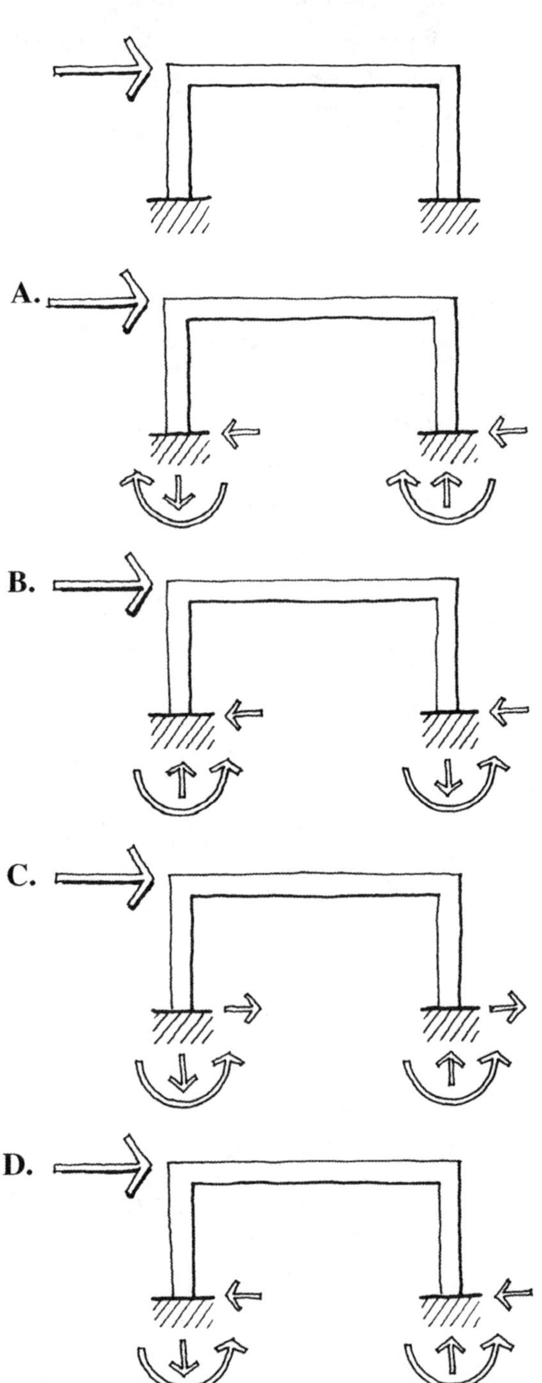

86. Several structural systems are being considered for a 20-story office building with columns 30 feet apart in each direction. Which of the following floor framing systems is *LEAST* likely to be economical?

A. Concrete fill over steel deck over steel composite floor beams spaced at 10 feet on center that span 30 feet between steel girders

B. Concrete on metal forms over open web joists spaced at 2 feet on center that span 30 feet between steel girders

C. Concrete topping over precast prestressed concrete planks that span 30 feet between steel girder

D. Flat plate spanning 30 feet in each direction directly to the columns

87. The vertical cylindrical tank shown is used to store water. The resulting stress in the tank wall is

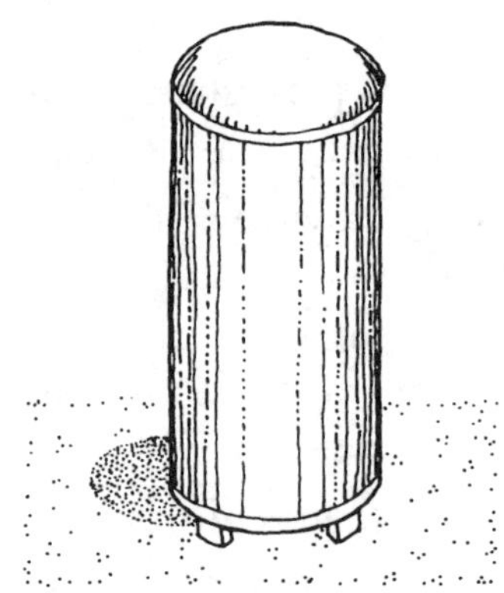

A. tension.

B. compression.

C. shear.

D. bending.

88. A beam fixed at both ends supports a concentrated load at the center of the span, as shown. Neglecting the weight of the beam, which of the following diagrams correctly shows the deflected shape of the beam?

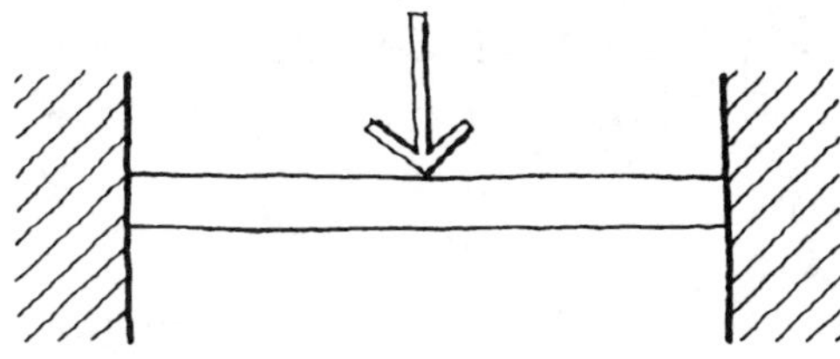

A. 

B. 

C. 

D. 

89. Complete the following statement. A three-hinged arch is

A. stiffer than a fixed arch.

B. statically determinate.

C. unstable when subject to horizontal loads.

D. generally semi-circular in profile.

90. In historic architecture, the transition from a round dome or drum to the pillars at the corners of a square space below was achieved by

A. a lantern.

B. a drum.

C. pendentives.

D. a vault.

91. Select the correct statements about the Leaning Tower of Pisa.

I.   It began to tilt during construction.

II. The leaning was caused by differential settlement of the underlying soil.

III. The top of the tower is out of plumb by 16 inches.

IV. The leaning was caused by the unbalanced structure of the tower.

A. II and III

B. I and IV

C. I, II, and III

D. I and II

92. Pier Luigi Nervi is associated with all of the following design techniques or materials, *EXCEPT*

A. ferrocement.

B. tubular frames.

C. concrete lamella roofs.

D. suspension roofs.

93. The moment diagram for a beam is shown below. Which of the following is the corresponding shear diagram?

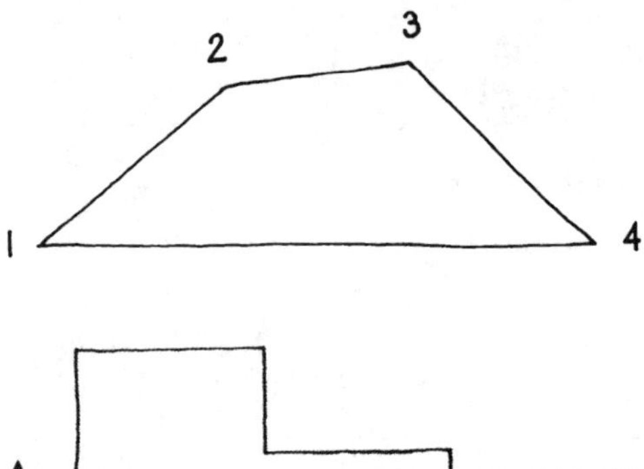

A.

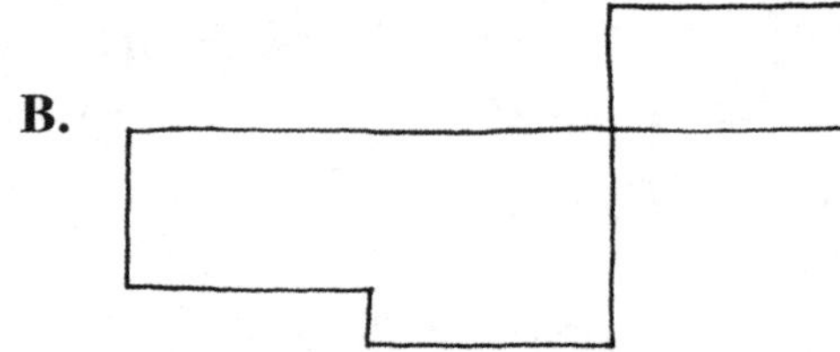

B.

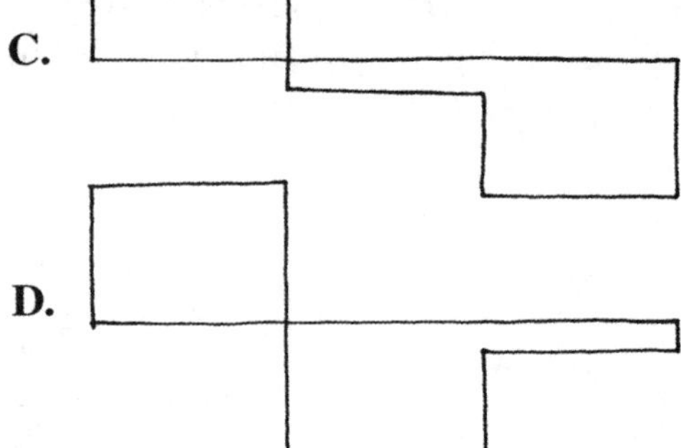

C.

D.

94. A structural steel beam spans 28 feet and supports a uniformly distributed load of 1,500 pounds per lineal foot. The maximum permissible deflection of the beam is 1 inch. What is the required I for the beam, if the value of E is 29,000,000 psi? Use the formula $\Delta = 5wL^4/384EI$.

A. 255 in.^{4}

B. 715 in.^{4}

C. 858 in.^{4}

D. 2,747 in.^{4}

95. Which of the following is the resultant of the two forces shown?

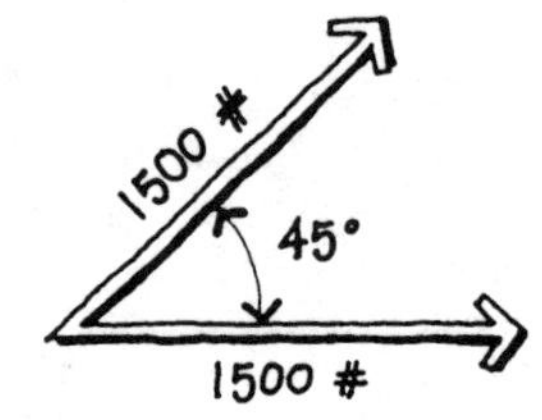

A.

B.

C.

D.

96. Which of the following is the best definition of live load?

   A. The load superimposed by the use and occupancy of a building, not including the wind, earthquake, or dead load.

   B. Any continuously applied load, except the dead load.

   C. The load superimposed by the use and occupancy of a building, including the wind or earthquake load, but not including the dead load.

   D. The weight of a building's occupants and movable furniture.

97. For which of the following building types does the structural cost represent the lowest percentage of the total cost of construction?

   A. Hospital

   B. Warehouse

   C. Parking garage

   D. Shopping mall

98. You are designing a parking garage with a span of 62 feet. Which of the following systems would probably be practical and economical?

   I. One-way concrete joist and beam

   II. Prestressed concrete tees

   III. Flat slab

   IV. Posttensioned concrete girders with pretensioned concrete planking

   A. I and III

   B. I and IV

   C. II and IV

   D. I, II, III, and IV

99. Which of the following correctly shows the reactions at A and B for the structure shown?

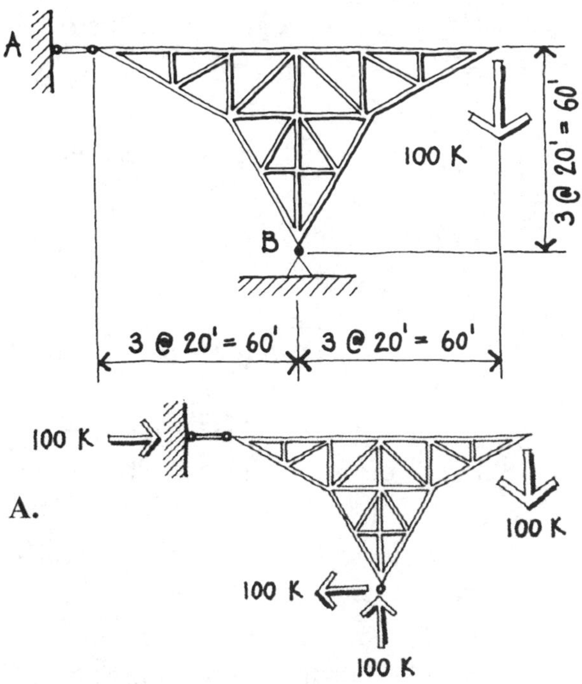

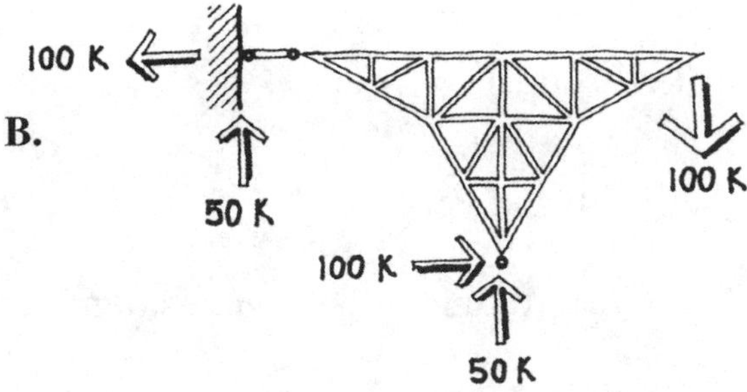

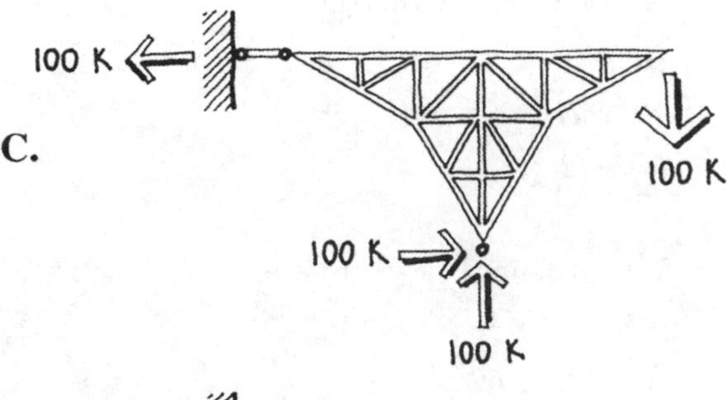

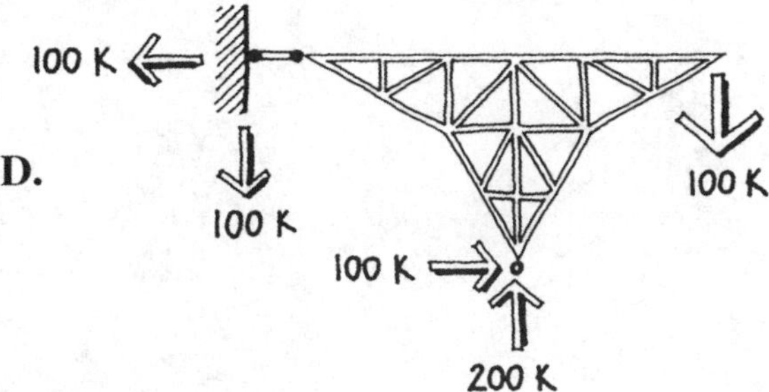

100. How can the calculated deflection of a steel beam be reduced?

    A. Use steel with a greater yield strength.

    B. Use a beam with a greater moment of inertia.

    C. Use a beam with a greater section modulus.

    D. Use steel with a greater modulus of elasticity.

101. Which of the following statements about cable structures is correct?

    I. When supporting loads that are uniformly distributed horizontally across its span, the shape of a cable is parabolic.

    II. When a cable is subject to changing loads, it changes its shape.

    III. The horizontal thrust at the ends of a cable is directly proportional to the sag of the cable.

    A. I and II

    B. II and III

    C. I and III

    D. I, II, and III

102. The ratio of unit stress to unit strain is called the

    A. moment of inertia.

    B. elastic limit.

    C. modulus of elasticity.

    D. yield point.

103. The stiffness of a member refers to its resistance to

    A. deformation.

    B. applied loads.

    C. impact.

    D. abrasion.

104. Comparing the two beams shown, the two-span continuous beam and the simple beam, which of the following statements is *INCORRECT*?

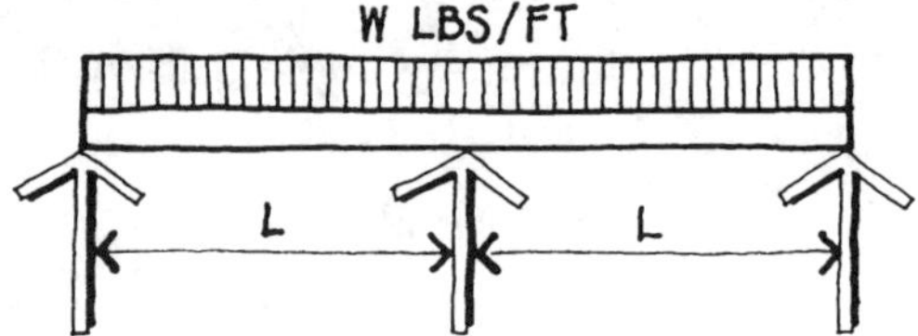

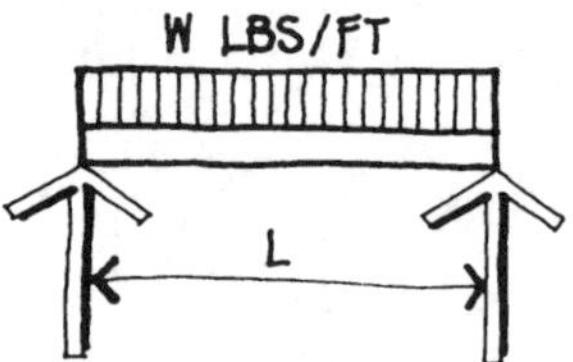

    A. The maximum positive moment in the simple beam is greater than in the continuous beam.

    B. The maximum shear in the simple beam is greater than in the continuous beam.

    C. The maximum deflection of the simple beam is greater than that of the continuous beam.

    D. The continuous beam has negative moment over the interior support, and the simple beam has no negative moment.

105. When a steel bar is subject to a tensile load, it increases in length. Up to a certain unit stress, the bar will return to its original length when the load is removed. What is this unit stress called?

    A. Modulus of elasticity

    B. Ultimate strength

    C. Yield point

    D. Elastic limit

106. The load capacity of a structural steel column depends on the ratio Kl/r. In this regard, which of the following statements are correct?

   I. K is a constant determined by the end conditions of the columns.

   II. l is the effective length of the column.

   III. r is the radius of gyration, which depends on the yield strength of the steel.

  A. I only

  B. I and II

  C. I and III

  D. II and III

107. What is the force in bolt A resulting from the handrail detail shown below?

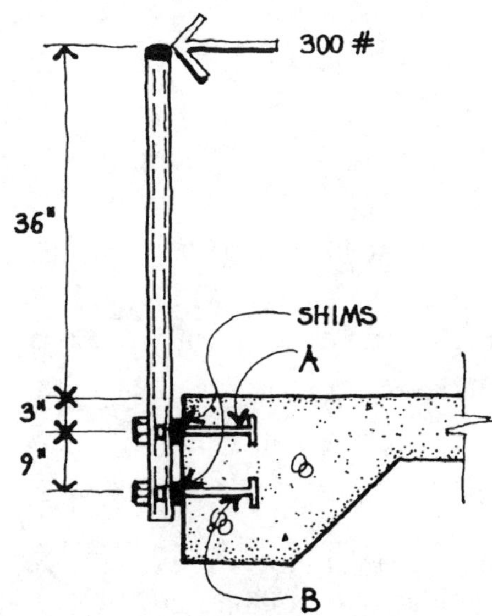

  A. 1,300 lbs. tension

  B. 1,600 lbs. tension

  C. 300 lbs. tension

  D. 1,300 lbs. compression

108. A steel bar two inches in diameter and 20 feet long resists a tensile load of 50,000 pounds. What is the unit tensile stress in the bar?

  A. 7,958 psi

  B. 12,500 psi

  C. 15,915 psi

  D. 25,000 psi

109. Which of the beams below are statically indeterminate?

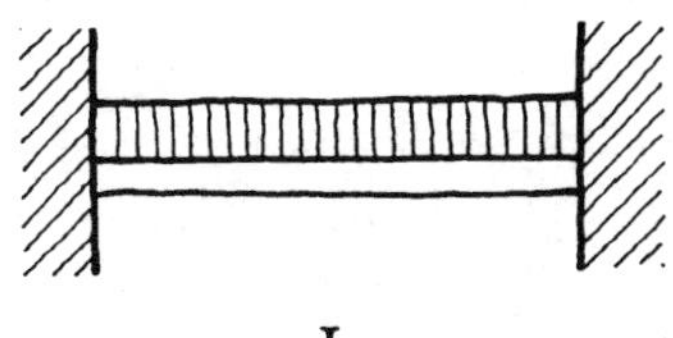

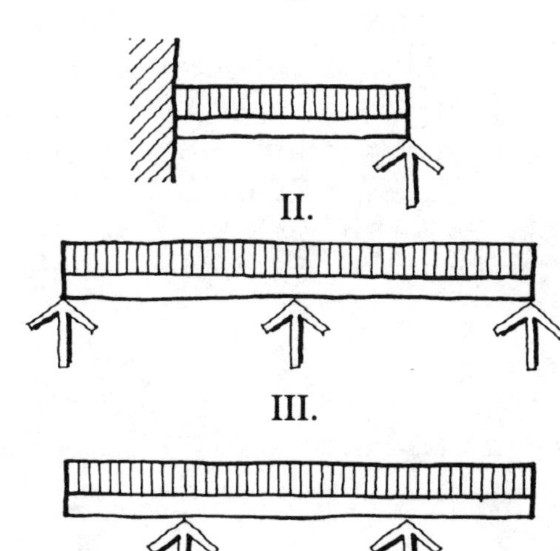

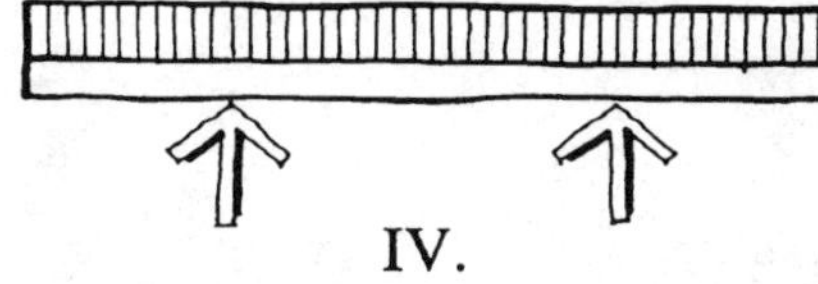

  A. I, II, and III

  B. I and III

  C. II and IV

  D. I and II

110. The roof of a circular building is supported by a series of radial cables, which connect to a ring at the center of the roof and to another ring at the perimeter. Which of the following statements is correct?

A. The center ring is higher than the perimeter ring.

B. Both rings are in tension.

C. The center ring is in tension and the perimeter ring is in compression.

D. Both rings are in compression.

111. A beam supports the loads shown below. What is the reaction at R_1, neglecting the weight of the beam?

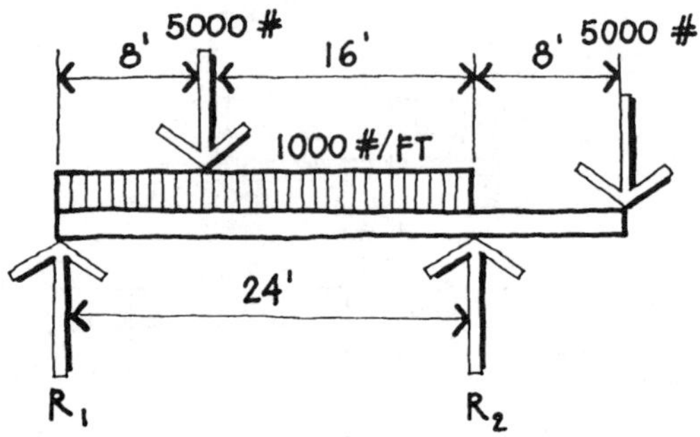

A. 20,333#

B. 17,000#

C. 15,333#

D. 13,667#

112. A bar two inches in diameter and 10 feet long stretches 0.159" when subject to a tensile load of 50 kips. What is the modulus of elasticity of the bar?

A. 1,001,000 psi

B. 10,010,000 psi

C. 12,012,000 psi

D. 29,000,000 psi

113. Which of the following factors are more critical for long span buildings than for conventional buildings?

I. Field inspection and testing

II. Snow drift loads and partial snow loads

III. The effects of temperature, creep, and shrinkage

IV. Secondary stresses caused by deflection and the interaction of building elements

A. I and III

B. II and IV

C. I, II, and IV

D. I, II, III, and IV

114. Which of the following statements about the continuous beam shown is *INCORRECT?* Neglect the weight of the beam.

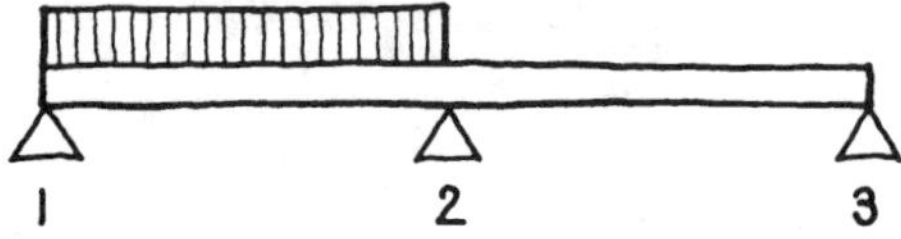

A. The moment in span 1-2 is always negative.

B. The moment over support 2 is negative.

C. The reaction at 3 is downward.

D. The moment in span 2-3 is always negative.

115. Which of the following statements concerning an object in equilibrium is *INCORRECT?*

A. There is no unbalanced force acting on the object.

B. The resultant force on the object passes through the centroid.

C. There is no unbalanced moment acting on the object.

D. The resultant force on the object is equal to zero.

116. A membrane inflated by air pressure encloses an occupied space and is anchored to a reinforced concrete ring at its perimeter. Which of the following statements is correct?

I.   The ring is in compression.

II. The ring is in tension.

III. The membrane tends to lift off the ring.

IV. The membrane tends to push down on the ring.

A. I and IV

B. I and III

C. II and IV

D. II and III

117. A tank is filled with water to a depth of ten feet. Which of the following diagrams correctly shows the pressure exerted on the tank walls? The unit weight of water is 62.4 pounds per cubic foot.

A.

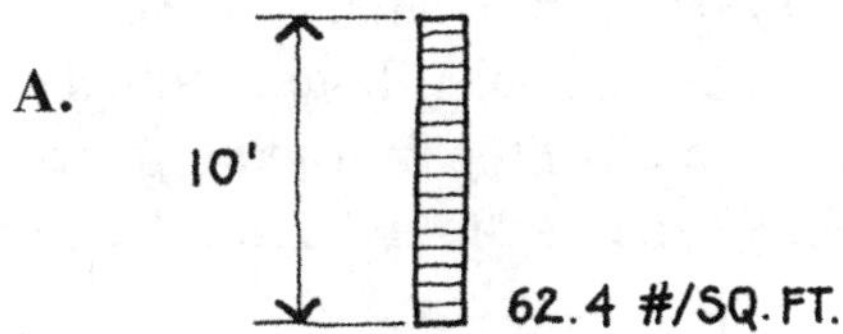

B.

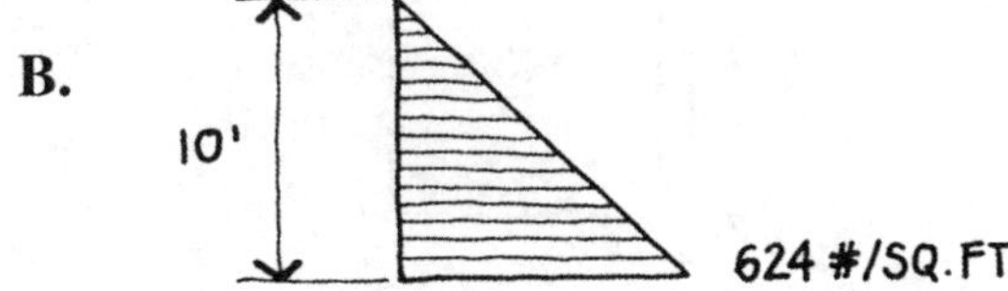

C.

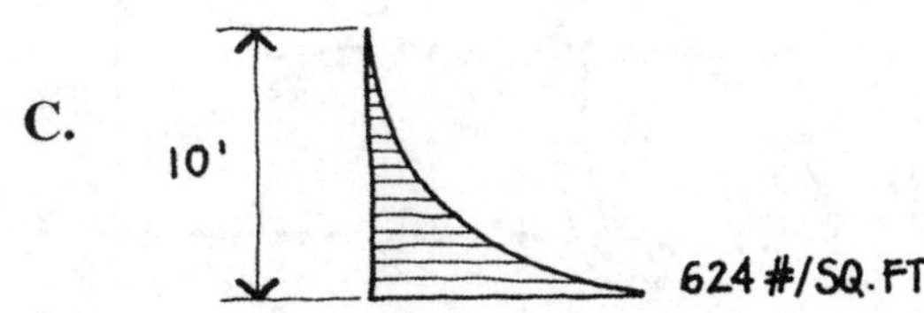

D.

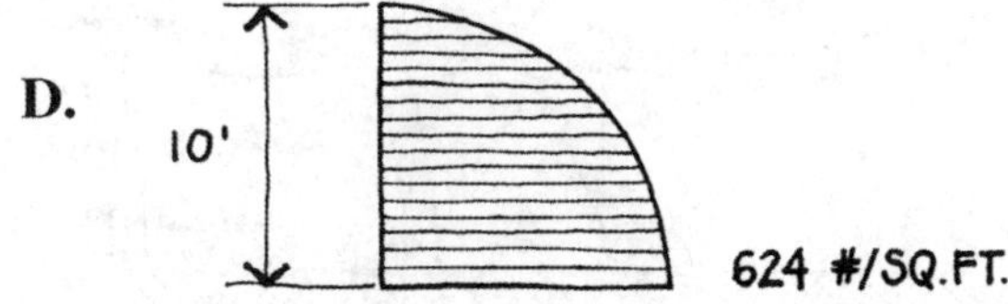

118. Which of the following diagrams correctly shows the reactions acting on the structure shown?

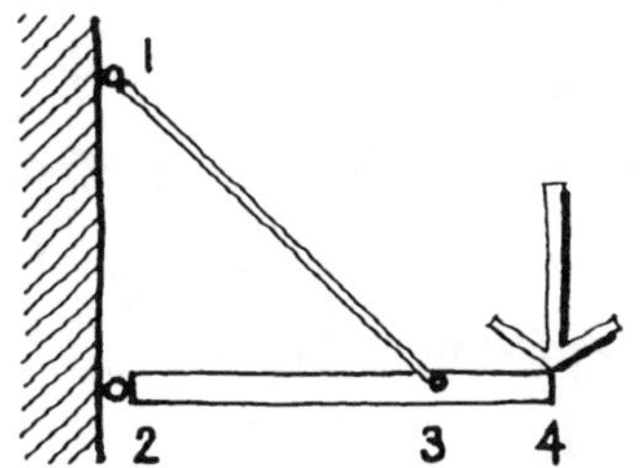

A.

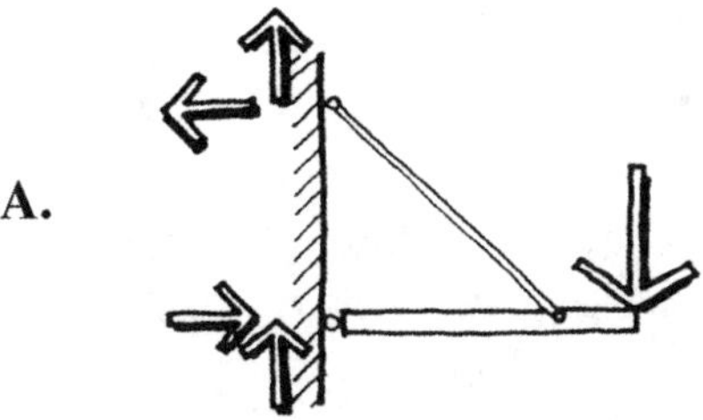

B.

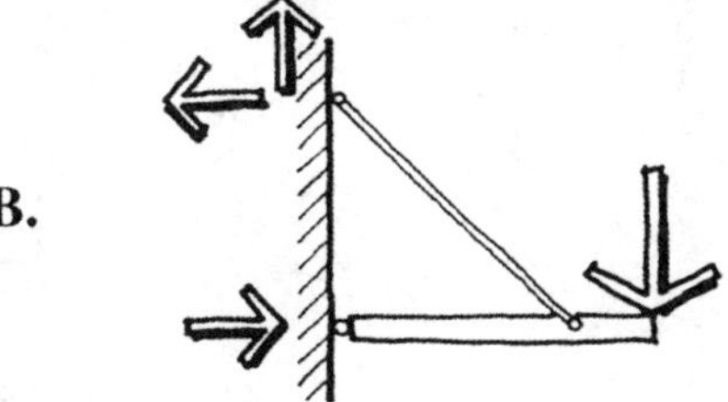

C.

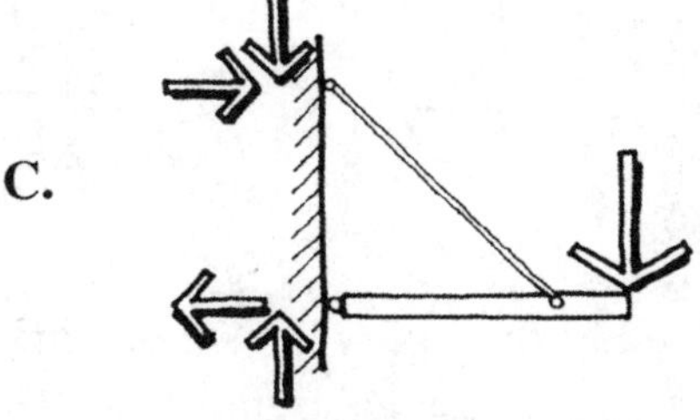

D.

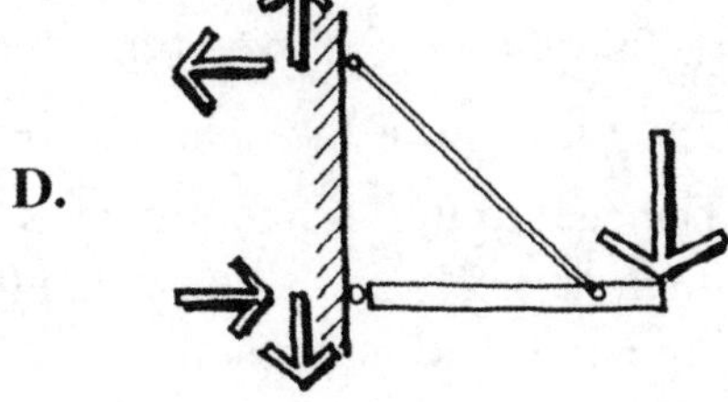

119. The change of length that a member undergoes when loaded axially depends on which of the following?

    I.   The magnitude of the load

    II. The length of the member

    III. The moment of inertia of the member

    IV. The cross-sectional area of the member

    V.   The modulus of elasticity of the material

    A. II, III, and IV

    B. I, II, IV, and V

    C. I, III, and V

    D. I, II, III, IV, and V

120. Which of the following correctly shows the variation of shear stress within the depth of a rectangular beam?

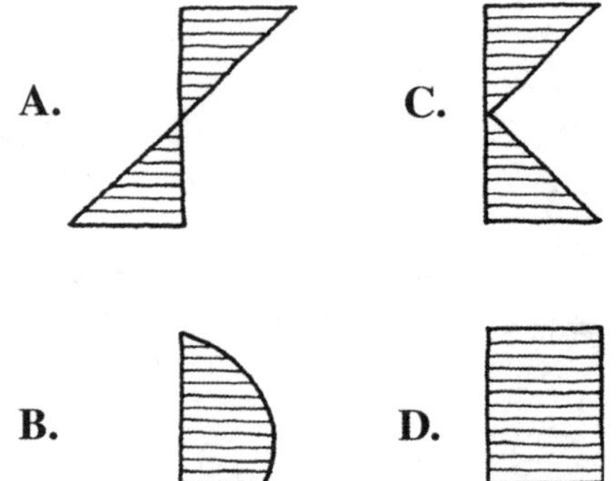

121. Which of the following statements about a three-hinged gabled frame is *INCORRECT?*

    A. It is statically determinate.

    B. Its supports permit rotation.

    C. The moment at the center is zero.

    D. The maximum moment occurs at the intersection of the column and the sloping beam and is generally greater than in a rectangular rigid frame.

122. In general, the internal forces in the web members of a parallel chord truss

    A. decrease toward the center of the span.

    B. increase toward the center of the span.

    C. remain relatively constant across the span.

    D. may increase or decrease toward the center of the span.

123. Which of the following statements is *INCORRECT?*

    A. The cost of connections is a significant factor in selecting structural steel systems.

    B. Fillet welds are usually more economical than full penetration welds.

    C. Shop connections are usually more economical than field connections.

    D. Welded connections are usually more economical than bolted connections.

124. A one-story steel rigid frame with fixed bases supports a uniformly distributed vertical load as shown. Which of the following correctly shows the shape of the deflected frame?

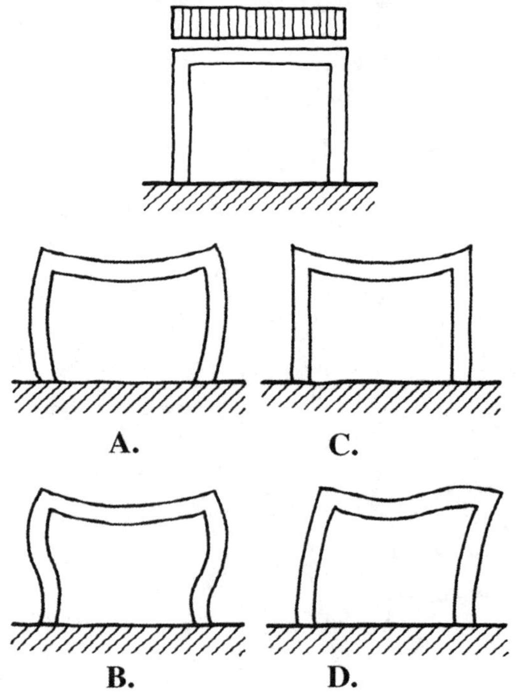

125. What is the moment about point O of the three forces shown above?

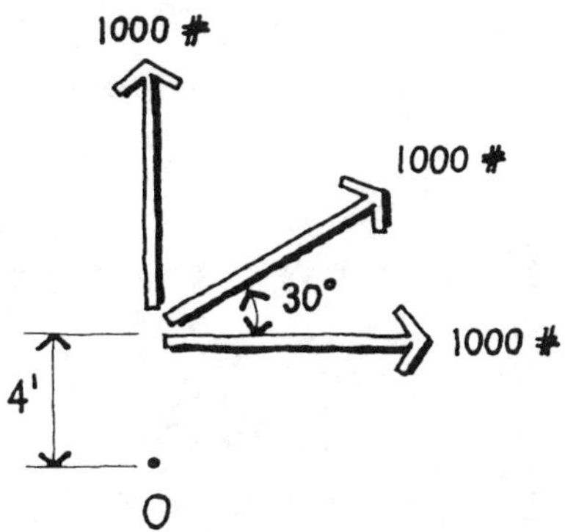

A. Zero

B. 4,000 ft.-lbs.

C. 7,464 ft.-lbs.

D. 12,000 ft.-lbs.

126. A 16-ft.-long steel column supports an axial load of 300 kips and is welded to a 20" × 20" base plate. What is the bearing pressure under the base plate?

A. 1,500 psi

B. 1,200 psi

C. 750 psi

D. 469 psi

127. A parabolic arch supports a uniformly distributed vertical load. What is the nature of the stress in the arch at any point along its length?

A. Pure compression, no bending moment

B. Pure bending moment, no compression

C. Combined compression and bending moment

D. Pure tension, no bending moment

128. Which of the live load arrangements shown below will result in the greatest positive moment in span 1-2?

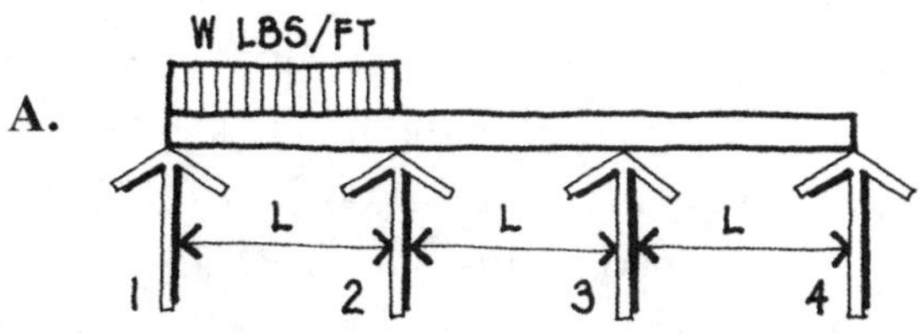

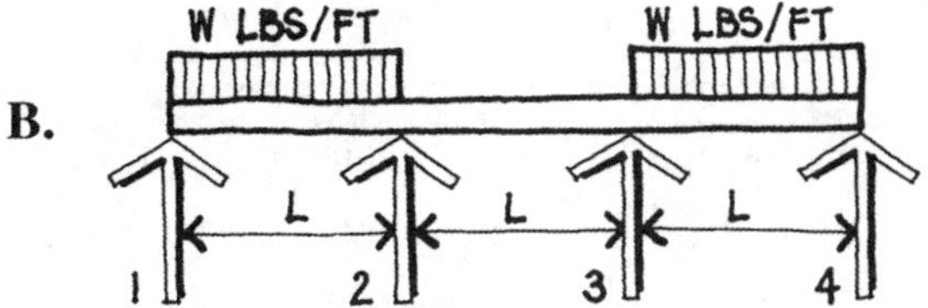

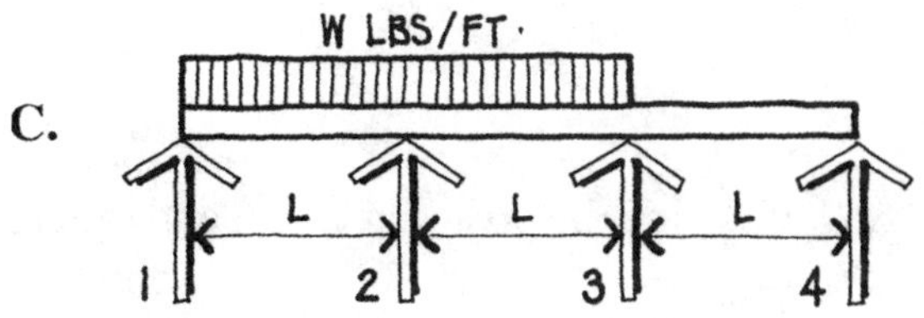

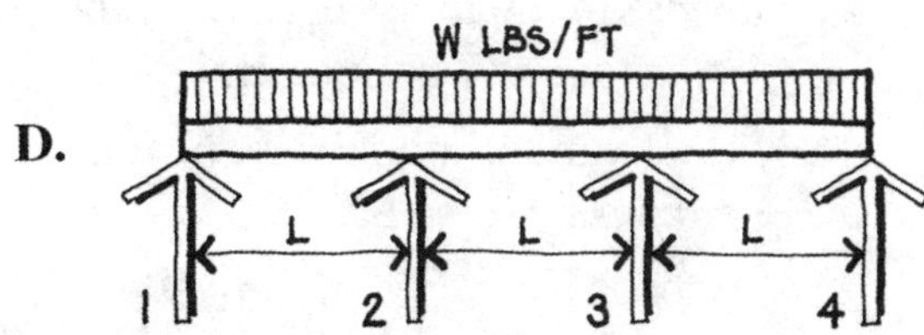

129. The radius of gyration r of a steel section is a function of the

I. modulus of elasticity.

II. moment of inertia of the section.

III. cross-sectional area of the section.

IV. section modulus of the section.

A. I and II

B. II and III

C. III and IV

D. I, II, and III

130. Which of the following statements concerning the economy of structural steel framing are correct?

   I. Moment connections should be used in preference to simple connections.

   II. Built-up sections should be used instead of rolled sections, to save weight.

   III. High-strength low-alloy steel should be used instead of ASTM A36 steel, because of its greater strength.

   IV. Composite steel deck should be used, rather than non-composite steel deck.

  A. III and IV

  B. I and III

  C. II and III

  D. IV only

131. Comparing the post-and-beam and the rigid frame shown below, which of the following statements are correct?

   I. The maximum moment in columns D is greater than in columns B.

   II. The maximum moment in beam C is less than in beam A.

   III. The axial force in columns D is the same as in columns B.

   IV. The horizontal reaction at the base of columns D is greater than at columns B.

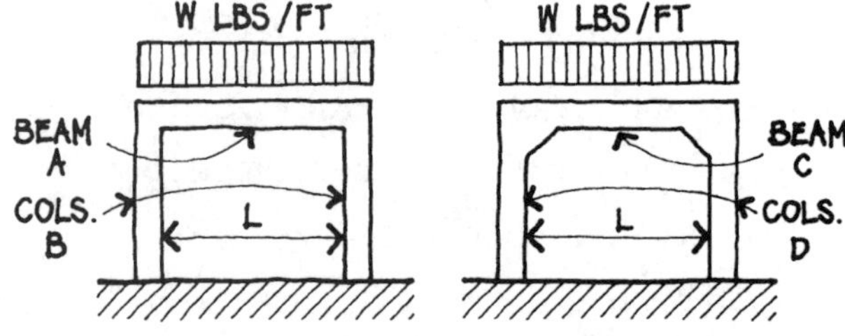

  A. I and II

  B. III and IV

  C. I and III

  D. I, II, III, and IV

132. What happens to a reinforced concrete dome when the exterior temperature rises?

  A. The top of the dome moves up.

  B. The top of the dome moves down.

  C. The dome develops internal compressive stresses.

  D. The dome develops internal tensile stresses.

133. Which of the structural steel framing plans below is usually the most economical?

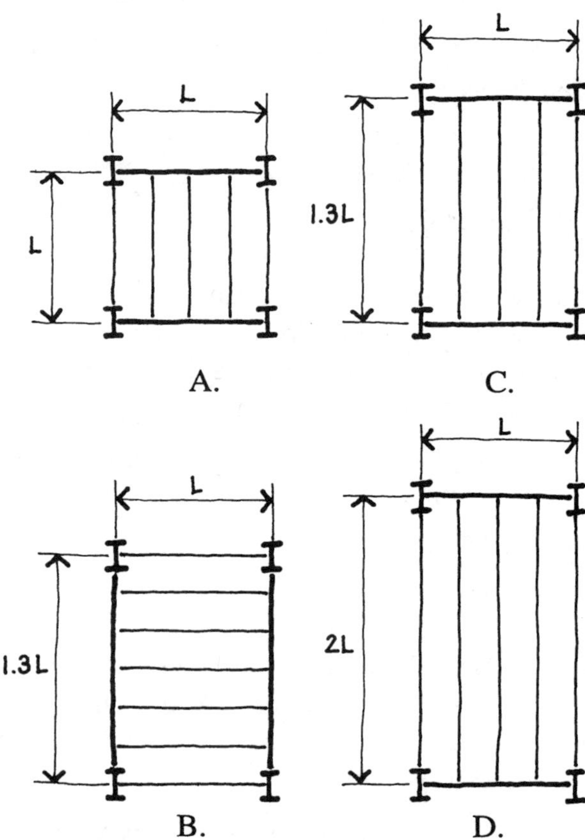

134. What is a rigid frame?

  A. A frame with diagonal braces

  B. A three-dimensional frame that supports a building's dead and live loads

  C. A steel or concrete frame that has bases fixed against rotation

  D. A frame with rigid joints capable of resisting moment

135. Which of the following systems provides maximum flexibility for a hospital without interruption of service?

A. Interstitial trusses

B. Staggered trusses

C. Space frame

D. Flat slab

136. Which of the following statements about Allowable Stress Design (ASD) and Load and Resistance Factor Design (LRFD) are correct?

I. ASD uses unfactored working loads and a single factor of safety.

II. LRFD uses separate factors for each load and for the resistance.

III. ASD provides uniform reliability for all steel structures under various loading conditions.

A. I and II

B. II and III

C. I and III

D. I, II, and III

137. Which of the following 19th-century buildings is considered to be the first true skyscraper?

A. Monadnock Building

B. Home Insurance Building

C. Carson Pirie Scott Store

D. Marshall Field Warehouse

138. All of the following are tubular steel buildings, *EXCEPT* the

A. Water Tower Building in Chicago.

B. Bank of China Tower in Hong Kong.

C. Sears Tower in Chicago.

D. John Hancock Building in Chicago.

139. The work of Pier Luigi Nervi is associated with which of the following?

I. Ferrocement

II. Prefabricated concrete domes

III. Prefabricated concrete lamella roofs

IV. Suspension bridge roof

A. II and III

B. I and IV

C. III and IV

D. I, II, III, and IV

140. The shape that a cable assumes when the only load acting on it is its own weight is called a

A. parabola.

B. catenary.

C. hyperbola.

D. paraboloid.

141. A thin shell structure is able to resist which of the following?

I. Shear

II. Tension

III. Compression

IV. Bending moment

A. II only

B. III only

C. I, II, and III

D. II, III, and IV

142. Section modulus is a measure of a beam's

A. resistance to deflection.

B. stiffness.

C. elasticity.

D. bending strength.

143. In the reinforced concrete footing shown, the minimum concrete coverage for reinforcement is

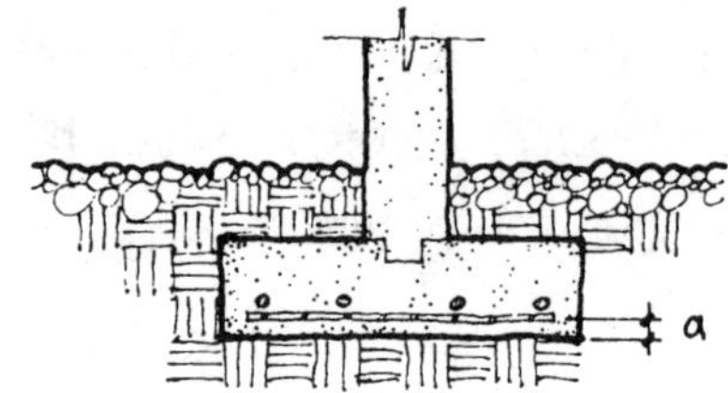

FOOTING

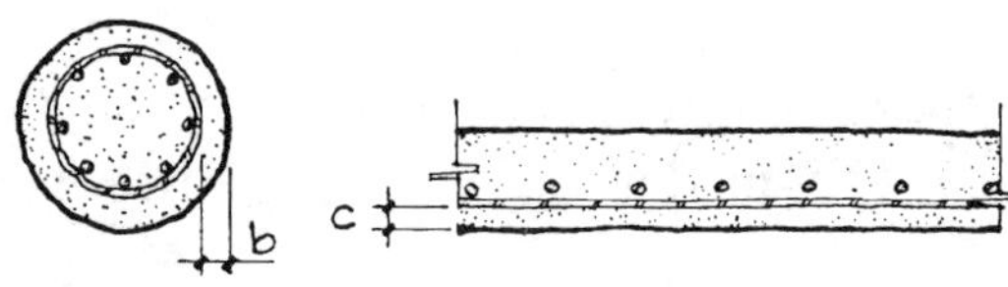

INTERIOR COLUMN   INTERIOR SLAB

    A. 1-1/2".

    B. 3".

    C. 6".

    D. 8".

144. Which of the following are *NOT* considered to be live load?

    I. Wind

    II. Earthquake

    III. Furniture

    IV. Snow

    A. II only

    B. I and IV

    C. I, II, and IV

    D. I, II, III, and IV

145. In the detail shown, the bolts conform to ASTM A325 in a bearing-type connection with threads excluded from the shear plane. Using Tables 1-D (p. 62) and 1-E (p. 63) from the AISC Manual of Steel Construction, what is the allowable load of the connection? Assume that the value of F_u of all connected material is

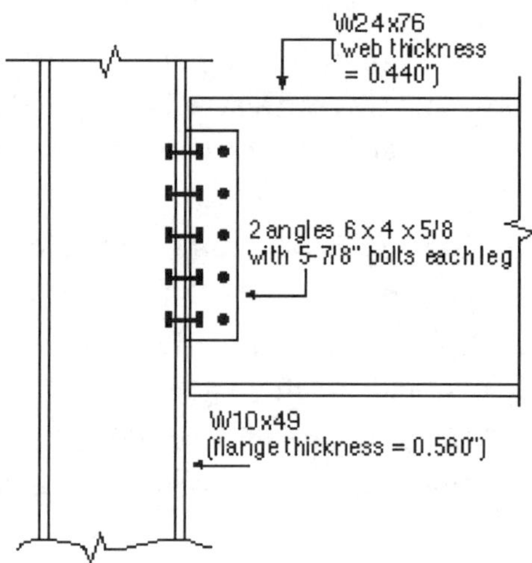

    A. 126.5 kips.

    B. 134 kips.

    C. 180.5 kips.

    D. 235.5 kips.

146. Select the correct statement concerning a change in the use or occupancy of an existing building.

    A. The building must comply with the live load requirements for the new use.

    B. The new use must be less hazardous than the existing use.

    C. The architect for the existing building must approve the change in use.

    D. The structural design of the building must be checked to verify that it will be safe for the new use.

147. The resistance to sliding of a retaining wall should equal at least

A. 75 percent of the total lateral pressure of the retained earth.

B. 90 percent of the total lateral pressure of the retained earth.

C. the total lateral pressure of the retained earth.

D. 1.5 times the total lateral pressure of the retained earth.

148. The basic stress in an arch is

A. compression.

B. tension.

C. flexure.

D. shear.

149. How are the allowable stresses for wood adjusted for the duration of the load?

A. The shorter the duration, the greater the allowable stresses.

B. The shorter the duration, the lower the allowable stresses.

C. The shorter the duration, the greater the allowable stresses for members, but not for mechanical fasteners.

D. The duration does not affect the allowable stresses.

150. Your design calculations for a retaining wall show that the dead load resisting moment is equal to 1.5 times the overturning moment from earth pressure. Which of the following statements is correct?

A. The design is satisfactory.

B. The footing should be made wider.

C. The footing should be made deeper.

D. The amount of reinforcing steel in the wall and footing should be increased.

151. Select the correct statements about Load and Resistance Factor Design (LRFD) for steel structures.

I.   The actual dead, live, and other loads are multiplied by their respective load factors.

II. The nominal strength is multiplied by a resistance factor ϕ, which is generally less than 1.

III. The actual unfactored loads are used in design.

IV. The factor of safety is obtained by using allowable stresses that are less than the yield stress.

A. I and II

B. III and IV

C. I and IV

D. II and III

152. Wide flange steel beams are efficient bending members because

A. they have thick webs.

B. most of their material is in the flanges.

C. they are symmetrical about both axes.

D. their strength is the same about both axes.

153. The footing shown in the plan is called a

A. mat foundation.

B. combined footing.

C. cantilever footing.

D. pile cap.

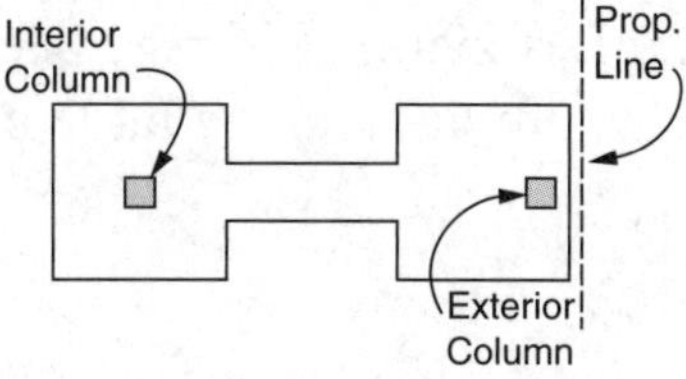

154. The main cables in a suspension-type roof structure are stressed in

    A. compression and bending.

    B. compression only.

    C. tension and bending.

    D. tension only.

155. The load carrying capacity of a square or rectangular wood column is determined by all of the following *EXCEPT* the

    A. radius of gyration of the column section.

    B. unbraced height of the column.

    C. species of lumber.

    D. dimensions of the column section.

156. Which of the following statements about composite design is *NOT* true?

    A. Composite design should only be used when conventional steel framing cannot support the design loads.

    B. Composite design is generally more economical than conventional steel framing.

    C. Deflection is more critical in composite design than in conventional steel framing.

    D. The shear connectors used in composite design can develop the ultimate capacity of the steel or the concrete, whichever is less.

157. A W24 × 68 steel beam supports a uniformly distributed load of 2,000 pounds per lineal foot on a simple span of 28 feet. How much does it deflect, if I for the beam is 1,830 in.^{4} and E is 29,000,000 psi? See the beam formulas provided in the table on page 65.

    A. 0.11"

    B. 0.19"

    C. 0.38"

    D. 0.52"

158. The strength of a fillet weld

    A. is equal to that of a complete penetration groove weld.

    B. is the same as that of the connected material.

    C. is based on the shear strength through the throat of the weld.

    D. is based on the tensile strength through the throat of the weld.

159. Reinforcing steel in the side opposite the tension side of a reinforced concrete beam does which of the following?

    I. Resists compression

    II. Resists shear stress

    III. Resists tension

    IV. Reduces the beam deflection resulting from creep

    A. I only

    B. I and III

    C. I and IV

    D. II, III, and IV

160. The purpose of the effective length factor K used in the design of structural steel columns is to

    A. evaluate the earthquake force that a column must resist.

    B. provide a method for designing columns with varying end restraints.

    C. modify the allowable column loads for different grades of steel.

    D. provide a method for designing columns that resist both moment and direct stress.

161. Which of the beam to column connections shown below might be used for a moment-resisting steel frame?

A. I and II

B. I and III

C. III only

D. I, II, III, and IV

162. What is the midspan deflection of the beam shown below? Neglect the weight of the beam. Assume $E = 29 \times 10^6$ psi and $I = 2,100$ inches[4]. See the beam formulas on page 65.

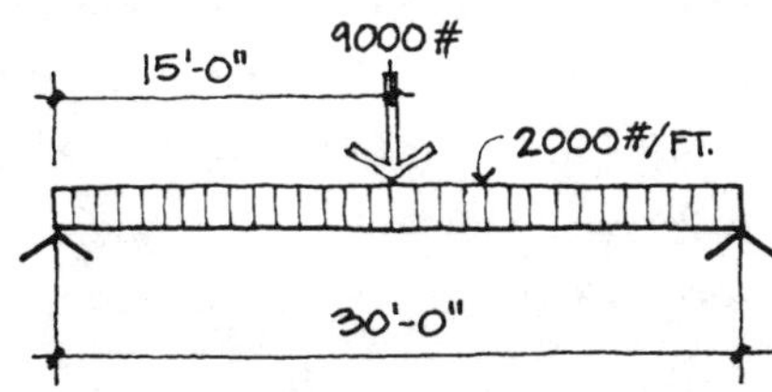

A. 0.346"

B. 0.498"

C. 0.599"

D. 0.743"

163. The calculated deflection of a single reinforced concrete beam is excessive, making it necessary to redesign the beam. Assuming no other structural elements are affected, and the load supported cannot be reduced, what is the most efficient way to reduce the beam's calculated deflection?

A. Make the beam wider.

B. Add reinforcing steel.

C. Specify higher strength concrete.

D. Make the beam deeper.

164. What is the maximum moment in the beam shown below?

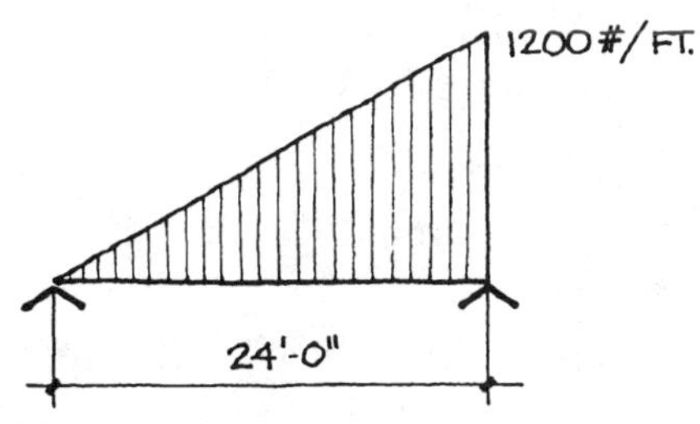

A. 88,681'#

B. 29,560'#

C. 44,340'#

D. 36,950'#

165. The nominal diameter and yield strength of the bar shown are as follows:

A. Diameter = 0.40 inch, yield strength = 60 ksi

B. Diameter = 1/2 inch, yield strength = 60 ksi

C. Diameter = 0.60 inch, yield strength = 40 ksi

D. The markings shown designate only the producer of the bar and the chemical content.

166. In the stress-strain diagram for steel shown below, what is A called?

- **A.** Elastic limit
- **B.** Modulus of elasticity
- **C.** Yield point
- **D.** Ultimate strength

167. In the stress-strain diagram for steel shown below, what is B called?

- **A.** Elastic limit
- **B.** Modulus of elasticity
- **C.** Yield point
- **D.** Ultimate strength

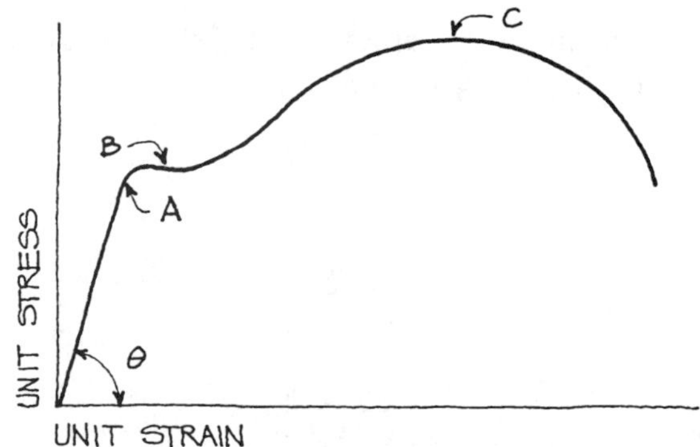

168. In the stress-strain diagram for steel shown above, what is C called?

- **A.** Elastic limit
- **B.** Modulus of elasticity
- **C.** Yield point
- **D.** Ultimate strength

169. In the stress-strain diagram for steel shown above, what is the tangent of angle θ called?

- **A.** Elastic limit
- **B.** Modulus of elasticity
- **C.** Yield point
- **D.** Ultimate strength

170. A building is spanned with 60 foot long 36LH07 steel joists placed at 6 feet on center. Using the table provided on page 64, how many lines of bridging are required for the joists?

- **A.** 4
- **B.** 5
- **C.** 6
- **D.** 7

171. Select the correct statements about trusses.

- I. Although trusses are efficient in supporting loads, they are deep and slender and therefore require adequate lateral bracing.
- II. When purlins are connected to trusses at the panel points, the top chord members are subject to axial stress only.
- III. Trusses are generally economical for spans between 60 and 200 feet.
- IV. The optimum depth to span ratio of a truss is about 1/24.
- V. Increasing the depth of a truss decreases the stress in the web members.

- **A.** I, II, and III
- **B.** I, III, and IV
- **C.** II, III, IV, and V
- **D.** I, II, and V

172. Select the correct statement.

- **A.** A trussed steel arch is generally used where the span is less than 100 feet.
- **B.** Steel arches are trussed in order to increase their bending resistance.
- **C.** Trussing a steel arch minimizes the horizontal thrust at the base.
- **D.** Using a trussed steel arch eliminates the need for a center hinge.

173. Long-span steel joists spanning 60 feet between girders are used to support a roof. The joists are spaced 6 feet on center, the dead load is 20 pounds per square foot (including the joist weight), the live load is 20 pounds per square foot, and the live load deflection is limited to 1/360 of the span. What is the lightest joist that can support the load? Use the table on page 64, which is reproduced from the Standard Specifications for Steel Joists.

A. 32LH06

B. 36LH07

C. 36LH08

D. 36LH09

174. A roof truss spans 80 feet and supports panel point loads of 15 kips each, as shown below. What is the maximum internal force in the top chord?

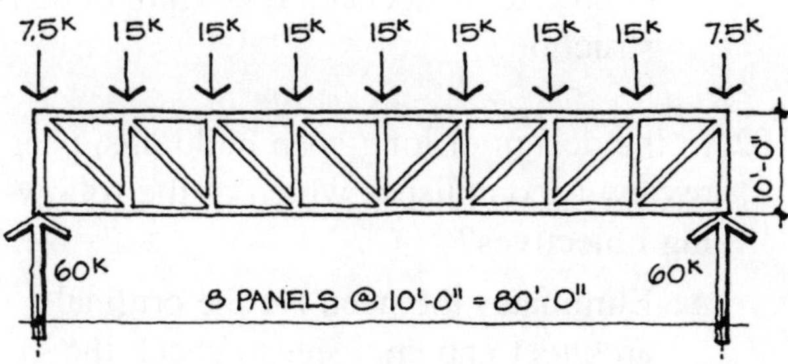

A. 15 kips

B. 60 kips

C. 112.5 kips

D. 120 kips

175. Reinforced concrete circular domes

A. generally resist flexural stresses.

B. are economical in the use of concrete.

C. are effective in resisting concentrated loads.

D. are inexpensive to form.

176. The roof of a building is supported by steel joists that span between joist girders. The typical bay is 60 feet by 60 feet, the joists are spaced at 6 feet on center, and the total dead and live load, including the joist girder weight, is 40 psf. What is the load at each panel point of a typical joist girder?

A. 7.2 kips

B. 12.0 kips

C. 14.4 kips

D. 24.0 kips

177. Which of the four space frame plan layouts shown below is most economical?

A.

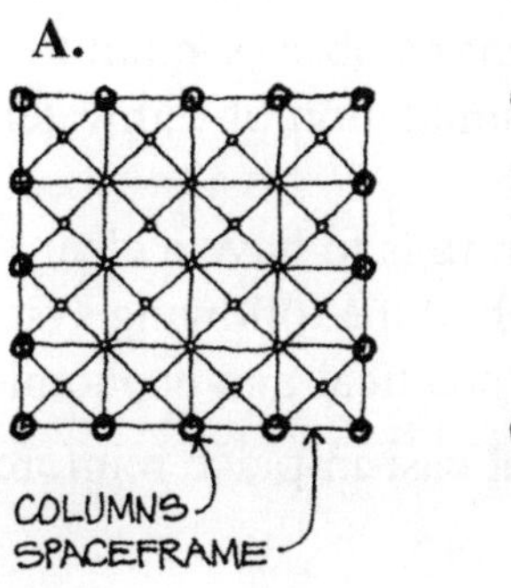

C.

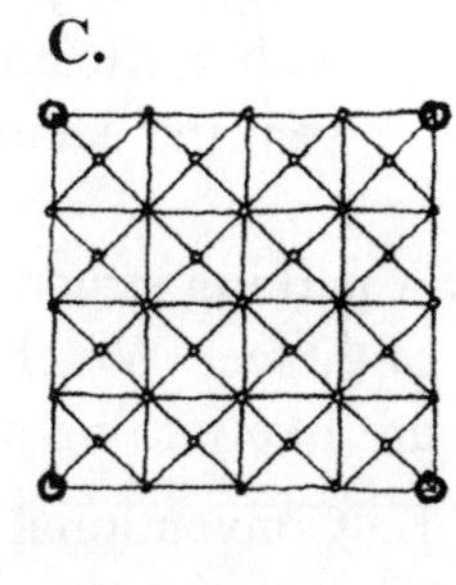

B.

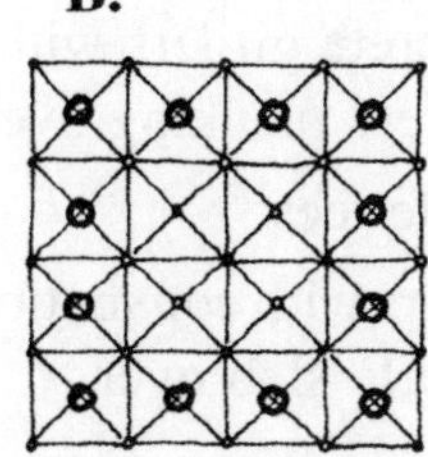

D.

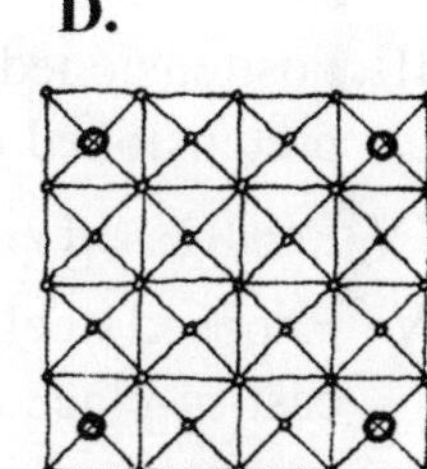

178. In the design of steel columns, the factor K is a function of

A. the length of the column.

B. the radius of gyration of the column.

C. the yield strength of the steel.

D. the end conditions of the column.

179. Which of the following statements about long-span open web steel joists is correct?

  A. Long-span joists generally have sufficient stiffness so that cambering the joists to prevent ponding of water is not necessary.

  B. The most economical way to prevent ponding of water is to build pitch into the top chords of the joists.

  C. Because of the possibility of ponding, long-span joists must be designed to support the weight of two inches of water uniformly over the entire roof surface.

  D. Long-span joists should be provided with sufficient camber or pitch to prevent the accumulation of rainwater.

180. A parking structure is to have a clear span of 62 feet. Which of the following systems are likely to be practical and economical?

  I.   Conventional cast-in-place reinforced concrete

  II. Glued laminated beams with heavy planking over

  III. Posttensioned concrete girders with pretensioned concrete planking over

  IV. Prestressed concrete tees

  V. Exposed steel girders with conventional cast-in-place concrete slab over

  A. All of the above

  B. IV and V

  C. III and IV

  D. III and V

181. You have selected a typical bay size 20 feet by 50 feet for a building for which you are the architect. The floor system consists of a concrete slab connected by studs to composite structural steel beams and girders. The beam spacing is 10 feet. You are investigating two different framing schemes: (1) 50 foot beams and 20 foot girders, and (2) 20 foot beams and 50 foot girders. Select the correct statement.

  A. Scheme 1 usually requires less weight of structural steel and less depth of construction.

  B. Scheme 2 usually requires less weight of structural steel and less depth of construction.

  C. Scheme 1 usually requires more weight of structural steel and less depth of construction.

  D. Scheme 2 usually requires more weight of structural steel and less depth of construction.

182. In the design of long-span buildings, peer review accomplishes which of the following objectives?

  A. Eliminates the need for the original architect and engineer to check the design

  B. Relieves the original architect and engineer of most responsibility for the design

  C. Does away with review by the local building department

  D. Provides an additional safeguard against design deficiencies

183. Select the correct statements about cable roof structures.

  I. A cable is always under tension.

  II. High strength cable is about four times as strong as structural steel.

  III. In a suspension roof, the cable material represents the largest part of the structural framing cost.

  IV. The dynamic behavior of cable-supported roofs is more critical than that of conventionally framed roofs.

  V. When supporting a load that is uniformly distributed horizontally across its entire span, a draped cable will assume the shape of a hyperbola.

A. III and V

B. I, III, and IV

C. I, II, IV, and V

D. I, II, and IV

184. What is the horizontal thrust at each end of the three-hinged arch shown below?

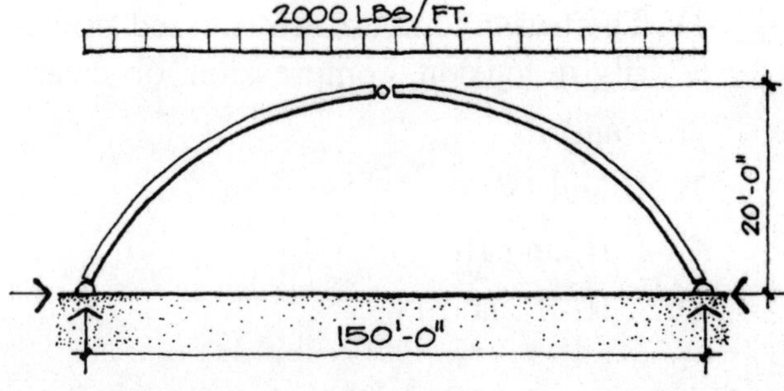

A. 75 kips

B. 150 kips

C. 281.25 kips

D. 562.5 kips

185. Select the correct statements about composite steel beam and concrete slab construction.

  I. Welded studs transfer moment between the steel beam and the concrete slab.

  II. For the same load and span, a composite beam is generally shallower than a noncomposite beam.

  III. A composite beam has greater stiffness than a noncomposite beam of equal depth, size, load, and span.

  IV. Deflection, particularly of the steel beam alone under construction loads, may be critical and should be calculated.

A. I and II

B. II and IV

C. I and III

D. II, III, and IV

186. Select the correct statements about reinforced concrete columns.

  I. A spiral column has more axial load-carrying capacity than an otherwise identical tied column.

  II. A spiral column has greater fire resistance than an otherwise identical tied column.

  III. Spiral columns are generally round, while tied columns are generally square.

  IV. Spiral columns have simpler forming than tied columns.

A. I only

B. I and III

C. II and III

D. I, II, III, and IV

187. A column has a tributary area of 500 square feet and supports a roof and one floor. The loads are as follows: Roof dead load—20 lbs. per square foot; Roof live load—20 lbs. per square foot; Floor dead load—75 lbs. per square foot; Floor live load—100 lbs. per square foot. Neglecting the weight of the column, what is the smallest pad footing which may be used, if the allowable soil bearing value is 3,000 pounds per square foot?

A. 4'-6" × 4'-6"

B. 5'-6" × 5'-6"

C. 6'-0" × 6'-0"

D. 7'-0" × 7'-0"

188. Select the correct statement about long span structures.

A. Long span structures generally have more redundancy than conventional structures.

B. Connections in conventional structures are more critical than those in long span structures.

C. Long span structures are more vulnerable to overall collapse than conventional structures.

D. Because of their inherent instability, long span structures generally require a dynamic load analysis.

189. Doubly reinforced concrete beams refer to

A. beams with twice the required reinforcing steel, in order to provide a greater factor of safety.

B. beams with compression as well as tension reinforcement.

C. beams with shear as well as tension reinforcement.

D. beams with tension reinforcement placed in two layers.

190. Which of the following statements about a circular dish roof are correct?

I.   The outer ring is in compression.

II. The outer ring is in tension.

III. The inner ring is in compression.

IV. The inner ring is in tension.

V.   The inner ring is lower than the outer ring.

VI. The inner ring is higher than the outer ring.

A. II, III, and V

B. II, III, and VI

C. I, IV, and V

D. I, IV, and VI

191. Which of the following statements about long span roof trusses are correct?

I.   Adequate lateral bracing should be provided.

II. Secondary stresses caused by joint restraint should be considered.

III. The trusses should be made up of triangles.

IV. The truss members are stressed primarily in tension, compression, or shear.

A. I and III

B. II and IV

C. I, II, and III

D. I, II, III, and IV

192. A beam of ASTM A36 steel supports a concentrated load of 10 kips at the center of a 30-foot span. Neglecting the weight of the beam, what is the required section modulus?

A. 75.0 in.^{3}

B. 37.5 in.^{3}

C. 25.0 in.^{3}

D. 18.75 in.^{3}

193. The roof of a convention hall has a column-free space that is 150 feet square. Two systems being considered are a space frame and a system of one-way parallel trusses. What are the advantages of the space frame over the parallel trusses?

I.   The space frame tends to deflect less.

II. The space frame can be made shallower.

III. Loads are resisted by all the members of the space frame.

IV. The space frame spans in two directions.

A. III and IV

B. II and IV

C. I and II

D. I, II, III, and IV

194. The deflection of a reinforced concrete beam is affected by which of the following?

I.   The load supported by the beam.

II. The duration of the load.

III. The span of the beam.

IV. The yield point of the reinforcing steel.

V.   The moment of inertia of the beam section.

VI. The strength of the concrete.

A. All of the above

B. I, III, and V

C. II, IV, and VI

D. I, II, III, V, and VI

195. What is the purpose of stirrups in reinforced concrete beams?

A. To resist shear stresses

B. To resist compressive stresses

C. To resist negative moment

D. To restrain the main reinforcement against buckling

196. What is the correct shear diagram for the beam shown?

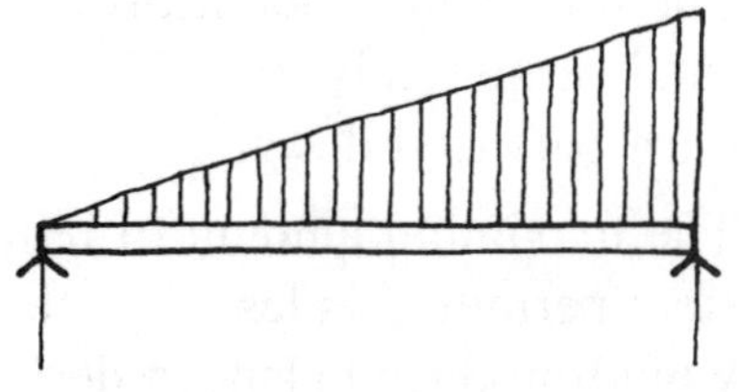

A. 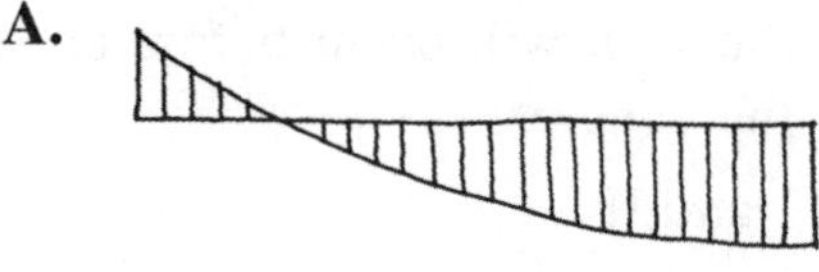

B. 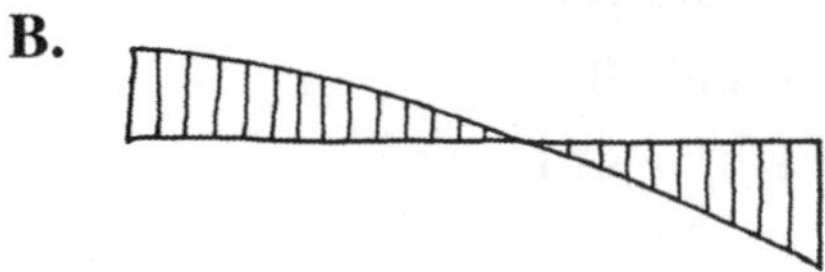

C. 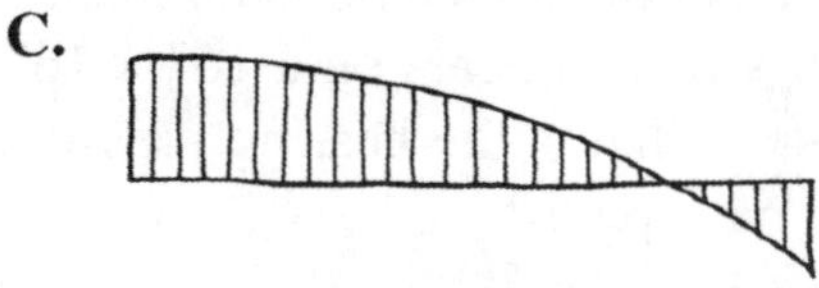

D. 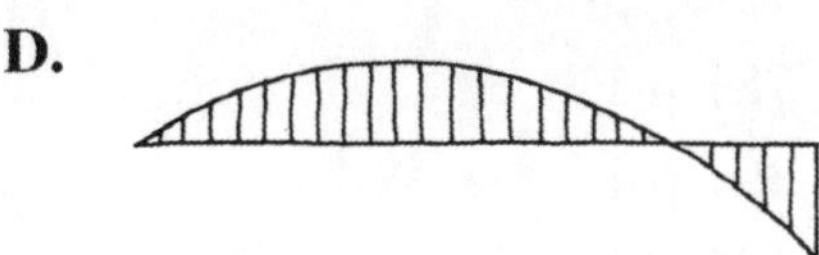

197. Complete the following statement. A rigid frame is

A. a system of beams and columns with diagonal bracing.

B. a structural steel frame in which all the joints are welded.

C. a frame restrained against rotation at its bases.

D. a frame with rigid joints that is able to resist vertical and lateral loads.

198. Select the correct statement concerning strength design of reinforced concrete.

   I. The minimum amount of reinforcing steel permitted is that which would produce balanced design.

   II. The maximum amount of reinforcing steel permitted is less than that which would produce balanced design.

   III. The design must assure that yielding of the steel will occur before crushing of the concrete.

   A. I and II

   B. II and III

   C. I and III

   D. I, II, and III

199. A steel column supports an axial load of 200 kips and bears on a 16" × 16" base plate. What is the bearing pressure under the base plate?

   A. 1,563 psi

   B. 781 psi

   C. 391 psi

   D. 125 psi

200. Which of the following methods of splicing reinforcing bars are allowed?

   I. Lapped splices

   II. Welded splices

   III. Mechanical connection splices

   A. I only

   B. I and III

   C. II and III

   D. I, II, and III

201. Which of the following statements about shell structures are correct?

   I. A shell structure can only resist compression in its own plane.

   II. Shell structures are suitable for resisting concentrated loads.

   III. Shell structures are too thin to resist any appreciable bending stresses.

   A. I and II

   B. II and III

   C. III only

   D. I, II, and III

202. The reaction at support C of the continuous beam shown below is 1.5 kips downward. Neglecting the weight of the beam, what is the reaction at support A?

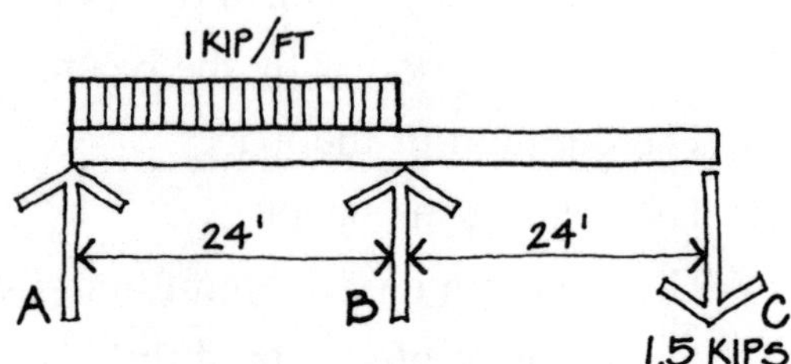

   A. 10.5 kips

   B. 12.0 kips

   C. 13.5 kips

   D. 15.0 kips

203. The floor framing of a building uses I-shaped wood joists spaced at 16 inches on center. The joists consist of solid wood flanges and a board web, which is inserted and glued into grooves in the flanges. What type of stress is resisted by the glue?

   A. Compression

   B. Tension

   C. Shear

   D. Bending

204. Which of the following statements about the shear stress in wood beams is *INCORRECT*?

    A. The strength of wood beams in shear perpendicular to the grain should always be checked.

    B. The shear stress parallel to the grain at any point in a beam is equal to the shear stress perpendicular to the grain.

    C. The allowable shear stress parallel to the grain is less than the allowable compressive or tensile stress parallel to the grain.

    D. A notch on the tension face at the end of a beam greatly increases the shear stress.

205. The modulus of elasticity of concrete increases as

    I.   its unit weight increases.

    II. its strength increases.

    III. the amount of reinforcing steel increases.

    IV. the depth of the member increases.

    A. I and II

    B. II and III

    C. III and IV

    D. I, II, III, and IV

206. What is concrete cover?

    A. The clear distance from the top of a concrete slab to the bottom reinforcing steel

    B. A covering placed over concrete during curing

    C. Protective wrapping over concrete members that are precast off-site

    D. The minimum thickness of concrete measured from the face of the concrete to the reinforcing steel

207. In the strength design method for reinforced concrete, the margin of safety is provided by two different factors: the load factor and ϕ, the strength reduction factor. In this regard, which of the following statements is *NOT TRUE?*

    A. The load factor for dead and live loads is always greater than 1.

    B. ϕ is always greater than 1.

    C. The load factor is based on the possibility that the service loads may be exceeded.

    D. ϕ allows for variations in materials, dimensions, and calculation approximations.

208. The base plate under a steel column spreads the column load over a large area of the supporting foundation. The base plate must be thick enough to resist the resulting

    A. compression.

    B. tension.

    C. bending.

    D. shear.

209. Complete the following statement. Open web steel joists

    A. are standardized lightweight steel trusses that may be shop or field-fabricated.

    B. have no fire resistance unless combined with other materials.

    C. are limited to spans of 60 feet.

    D. are generally designed as rigid frames.

210. In comparing the two trusses shown below, which of the following statements is correct?

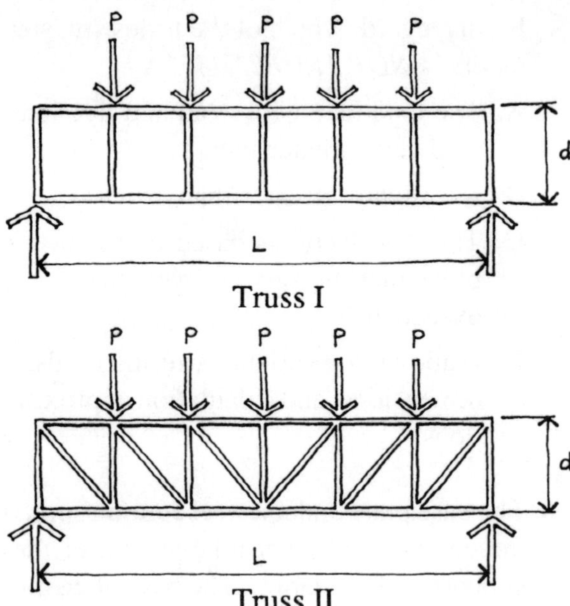

A. The chords in truss I have greater axial tension and compression than those in truss II.

B. Because of the absence of diagonals, truss I is unstable.

C. The chords and verticals of truss I are subject to bending stress.

D. Truss II will deflect more than truss I.

211. Which of the following are advantages of Vierendeel trusses?

I.   Unencumbered webs

II. Simple joints

III. Low deflection

IV. Economy

A. I and II

B. II, III, and IV

C. I and III

D. I, II, III, and IV

212. In the glued laminated roof system shown, what is the main purpose of connections A and B?

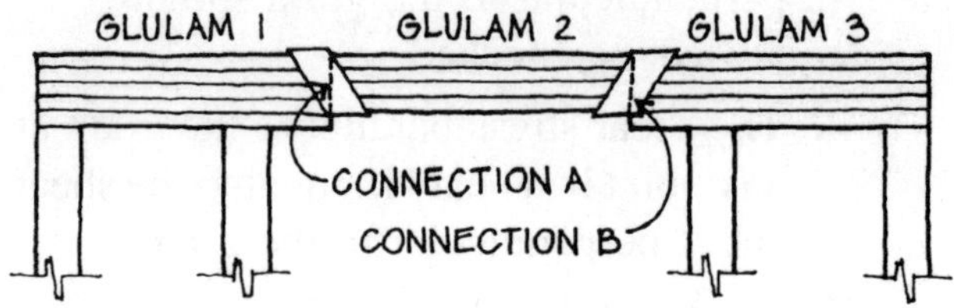

A. To transfer the vertical reactions of glulam 2 to glulams 1 and 3

B. To transfer the vertical reactions of glulams 1 and 3 to glulam 2

C. To provide a moment splice between the glulams

D. To transfer the horizontal reactions of glulam 2 to glulams 1 and 3

213. A steel beam supports a load of 1,000 pounds per foot on a simple span of 24 feet. The deflection is not to exceed L/360. What is the minimum required moment of inertia? $E = 29 \times 10^6$ psi and $\Delta = 5wL^4/384EI$.

A. 321.8 in.^{4}

B. 1,609 in.^{4}

C. 2,234 in.^{4}

D. 3,218 in.^{4}

214. What is a joist girder?

A. A steel plate girder used to support open web steel joists

B. A wide flange member that acts as a joist

C. An open web steel joist that is part of a rigid frame

D. A steel truss supporting equally spaced concentrated loads from a floor or roof system

215. Two trusses having the same load and span are shown. Which of the following statements are true?

I. The axial tension and compression forces in the chords are greater in truss A than in truss B.

II. The diagonals in truss B have greater axial tension forces than those in truss A.

III. The verticals in truss A have the same axial compression forces as those in truss B.

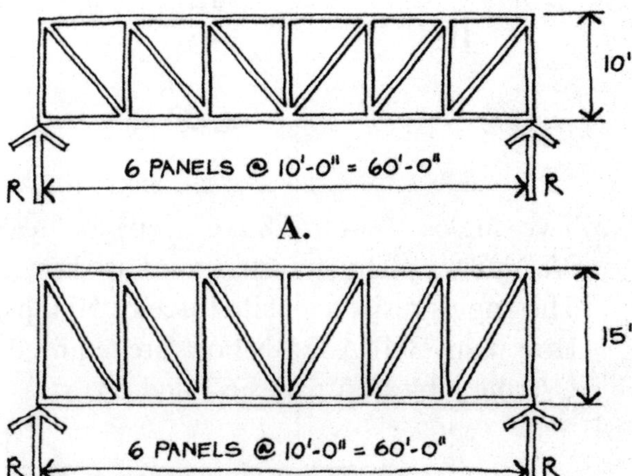

A. I and II

B. II and III

C. I and III

D. I, II, and III

216. A two-story warehouse has 40-foot by 40-foot bays. Which of the following reinforced concrete systems is likely to be the most economical?

A. Flat plate

B. Flat slab

C. Waffle slab

D. Pan joist

217. Which of the following statements about prestressed concrete is *INCORRECT*?

A. Prestressed members have no tension cracks because they are completely in compression.

B. Prestressed concrete has greater material and labor costs than conventional reinforced concrete.

C. Posttensioned members are usually produced at a casting yard away from the building site.

D. Smaller sizes may be used with prestressed concrete, because the entire section is effective in resisting applied loads.

218. The maximum deflection of the cantilevered steel beam shown is limited to 0.67 inch. What is the minimum required moment of inertia? $E = 29 \times 10^6$ psi. Use the beam diagrams on page 69 and neglect the weight of the beam.

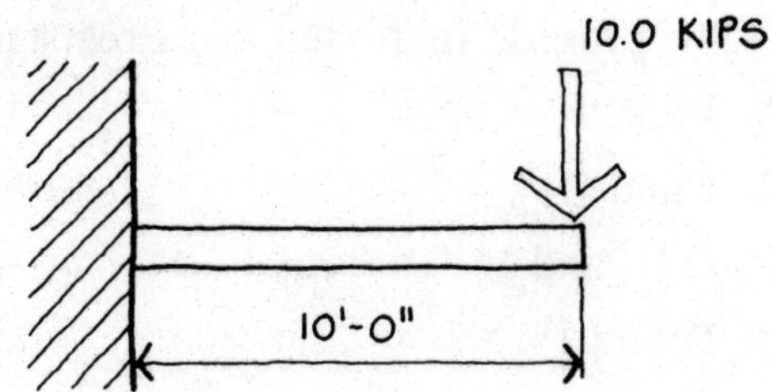

A. 24.7 in.^{4}

B. 171.6 in.^{4}

C. 205.9 in.^{4}

D. 296.4 in.^{4}

219. Which of the following are advantages of the stub girder framing system?

    I. Reduced weight of steel

    II. Reduced story height

    III. Simplified steel erection

    IV. Ease of providing simulated continuity for floor beams

A. I and II

B. II and III

C. I and III

D. I, II, III, and IV

220. Why are drop panels and column capitals used in flat slab construction?

    I. To reduce shear stress in the slab near the columns

    II. To provide greater effective depth for negative bending moment

    III. To provide greater effective depth for positive bending moment

    IV. To provide increased bond resistance.

A. I and II

B. I and III

C. I, II, and IV

D. II and IV

221. The staggered truss system is usually economical for

    I. spans greater than 45 feet.

    II. buildings under 8 stories in height.

    III. hotel and residential occupancies.

A. I and II

B. II and III

C. I and III

D. I, II, and III

222. Which of the four steel column sections shown below can support the greatest amount of vertical load, assuming they have the same cross-sectional area?

A.

B.

C.

D.

223. Two angles 6 × 4 × 3/8 are connected to a 3/8" gusset plate at a truss joint as shown. The angles resist a tensile force of 50 kips. How many 3/4" A325-N bolts are required? Use the tables on pages 62 and 63.

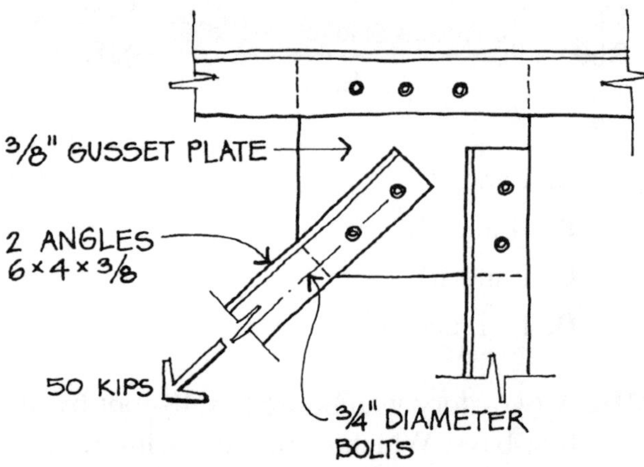

A. 2

B. 3

C. 4

D. 5

224. A beam of A36 steel with a span of 24 feet is installed at 40°F. It has unyielding supports. If the temperature rises to 100°F, what is the resulting stress in the beam? $E = 29,000$ ksi and the coefficient of expansion is 6.5×10^{-6}.

A. 11,310 psi compression

B. 11,310 psi tension

C. 13,572 psi compression

D. 18,850 psi compression

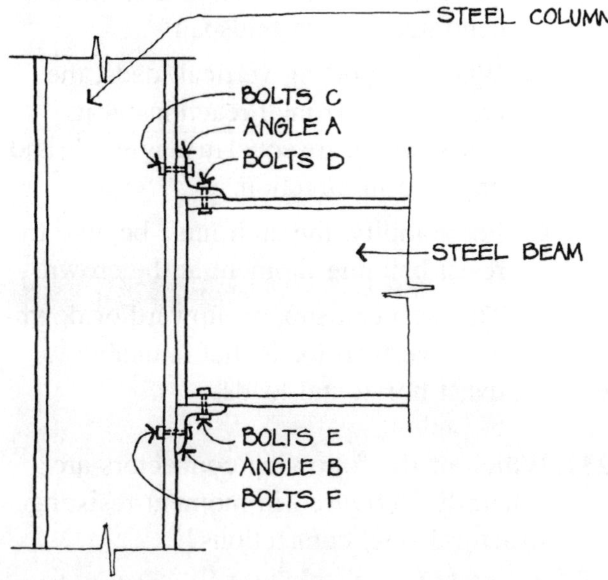

225. In the beam connection shown above, the beam reaction is transferred to the column through

A. angle A and bolts C and D.

B. angle B and bolts F.

C. angles A and B and bolts C and F.

D. angle B and bolts E.

226. In the connection shown,

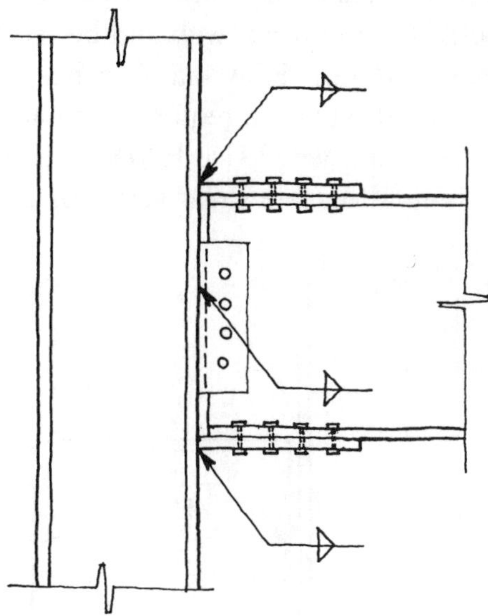

A. the web plate resists the shear and the flange plates resist the moment.

B. the web plate resists the moment and the flange plates resist the shear.

C. the web plate and flange plates resist the shear and there is no moment resistance.

D. the web plate and flange plates resist both the shear and moment.

227. In general, which of the following will help reduce the cost of structural steel framing?

I. Use moment connections in preference to non-moment connections.

II. Use rolled sections rather than built-up sections.

III. Restrict the number of members to a practical minimum.

IV. Use high-strength steel for all members.

A. I and III

B. II and III

C. II and IV

D. I, II, III, and IV

228. The beam shown has a reaction of 80 kips. It is attached to the column by an end plate welded to the beam web and bolted to the column flange. How many inches of 1/4 inch fillet weld are required on each side of the beam web? Use E70XX electrodes with an allowable shear stress of 21.0 ksi.

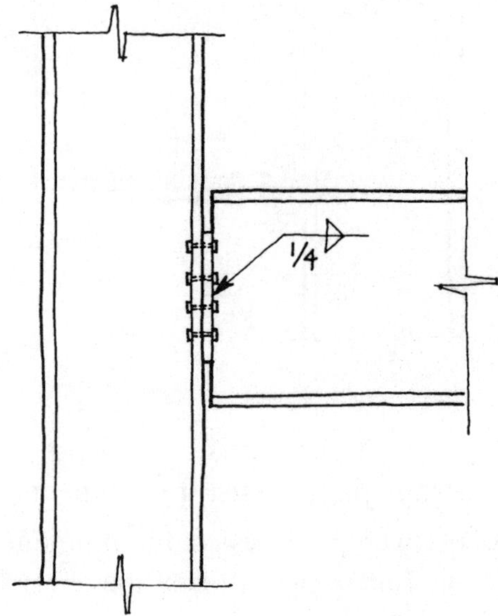

    A. 8"

    B. 11"

    C. 16"

    D. 22"

229. Select the correct statement about reinforced concrete columns.

    A. Reinforced concrete columns are normally designed for axial compression only.

    B. Spiral columns have greater axial load capacity than tied columns.

    C. Spiral and tied columns are both designed with the same strength reduction factor ϕ.

    D. Tied columns are more expensive than spiral columns.

230. Select the correct statement about the three-hinged arch shown below.

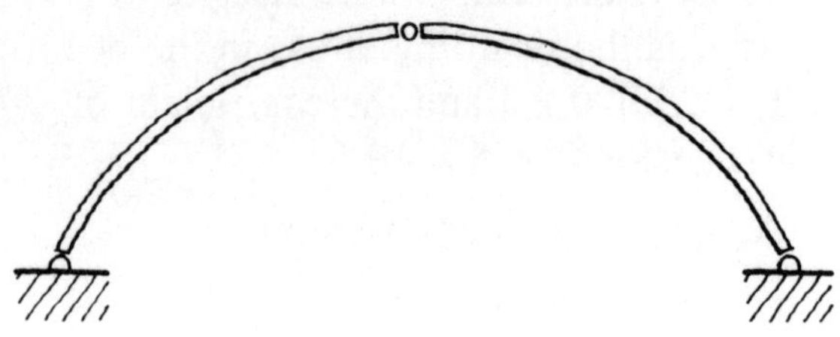

    A. There is no horizontal reaction at the bases if the arch supports only one concentrated load at midspan.

    B. When supporting vertical loads, the arch has horizontal reactions at its bases, which are equal in magnitude and opposite in direction.

    C. For stability, the arch must be able to resist bending moment at the crown.

    D. The arch can support upward or downward vertical loads, but is unable to resist horizontal loads.

231. Which of the following connectors are generally *NOT* used in moment-resisting structural steel connections?

    A. ASTM A325-SC high-strength bolts

    B. ASTM A307 machine bolts

    C. Full penetration groove welds

    D. Fillet welds

232. A 24-foot-long steel beam is installed when the temperature is 60°F. How much will it expand if the temperature rises to 90°F? The coefficient of expansion of steel is 0.0000065.

    A. 0.047"

    B. 0.056"

    C. 0.112"

    D. 0.168"

233. What is the principal purpose of a base plate under a steel column?

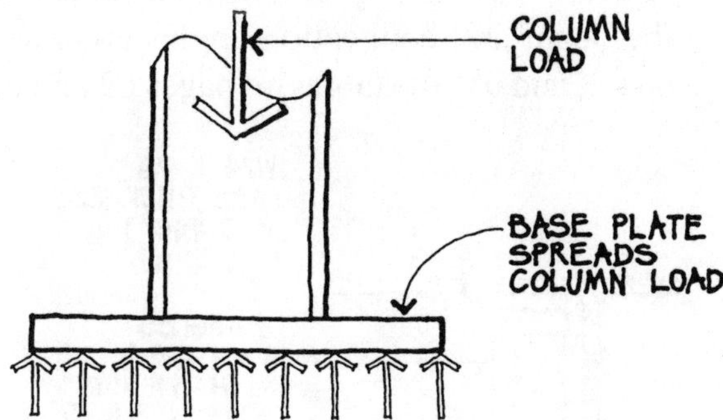

A. To act as a template for the anchor bolts

B. To provide fixity for the column

C. To provide a level surface for connecting the column

D. To spread the column load so that the bearing pressure on the foundation is not excessive

234. Which of the following are advantages of a conventional truss over a Vierendeel truss?

I. Less deflection

II. Less material

III. Less conflict between truss members and doors or windows

IV. Ability to have loads applied between panel points

A. I and II

B. I and IV

C. II and III

D. I, II, and IV

235. In a structural steel rigid frame, moment resistance at the beam-column joints may be provided by which of the following?

I. Two clip angles bolted to the beam web and the column flange

II. Groove welds between the top and bottom beam flanges and the column flange

II. Plates welded to the column flange and bolted to the top and bottom beam flanges

IV. A seat angle welded to the column flange and bolted to the bottom flange of the beam

A. I and II

B. I and III

C. I, II, III, and IV

D. II and III

236. A wood column is used to support an axial load of 40 kips. The column is 16 feet long, and its ends are restrained against lateral movement. The allowable stress in compression parallel to the grain is 0.30 $E/(l/d)^2$, where E = 1,600,000 psi, 1 = unsupported length of column, and d = least lateral dimension of column, but such stress may not exceed 1,150 psi. What is the smallest column that may be used?

A. 6×6

B. 6×8

C. 8×8

D. 8×10

237. The structural system shown below is called a

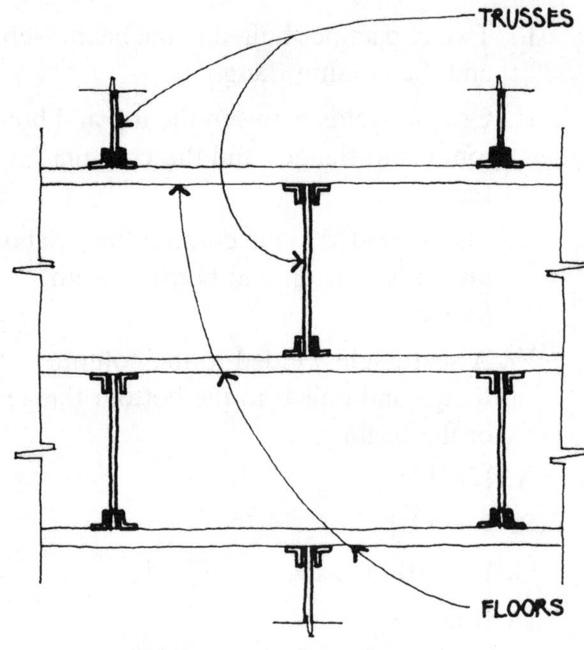

A. Vierendeel truss system.

B. staggered truss system.

C. space frame.

D. two-way truss system.

238. Which of the following statements about steel columns is correct?

A. The buckling tendency of a steel column depends on its yield point.

B. The maximum allowable slenderness ratio is 50.

C. Tubular sections have less tendency to buckle than wide flange sections.

D. If the value of r is different in each direction, the greater value is used to compute the Kl/r ratio.

239. What is the capacity of the connection shown? All connected material is ASTM A36 ($F_y = 36$ ksi, $F_u = 58$ ksi). Assume that the beam and connection angles are adequate, and use the tables on pages 62 and 63.

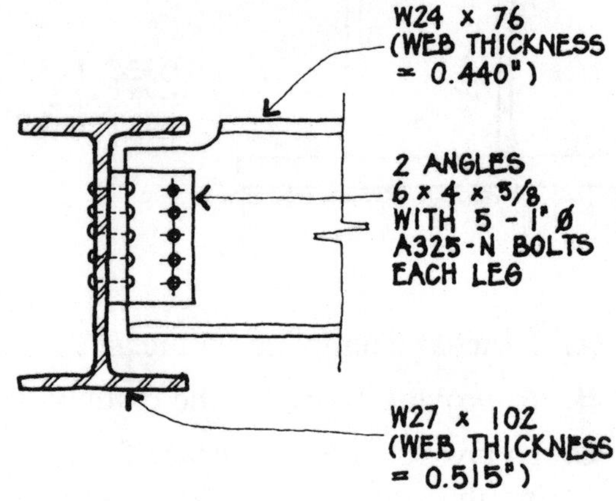

A. 82.5 kips

B. 153 kips

C. 165 kips

D. 330 kips

240. For the truss shown below, what is the internal axial force in member *a*?

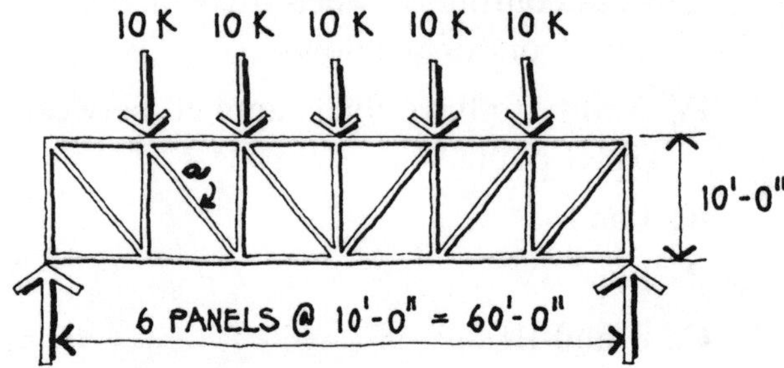

A. 10.6 kips tension

B. 14.1 kips tension

C. 15.0 kips tension

D. 21.2 kips tension

241. You are the architect for a gymnasium that is 120 feet by 120 feet in plan and has no interior columns. Two systems are being considered: a space frame and a system of one-way parallel trusses spaced 30 feet apart. Which of the following are advantages of the space frame over the parallel trusses?

I. Economy through the use of many identical members and connections

II. Greater stiffness

III. Less depth

IV. Simpler analysis

A. II and III

B. I and III

C. I, II, and IV

D. I, II, and III

242. What is the purpose of belling a caisson?

A. To increase the caisson's frictional resistance

B. To stabilize the soil around the caisson

C. To increase the caisson's bearing capacity

D. To make the excavation watertight

243. Which of the following factors affects the shear capacity of a reinforced concrete beam?

A. The cross-sectional area of the longitudinal tension reinforcing

B. The span of the beam

C. The ultimate 28-day strength of the concrete

D. The load on the beam

244. What is the maximum bending moment that can be resisted by a 6×12 wood beam, if $F_b = 1,600$ psi, $A = 63.25$ in.^{2}, $S = 121.229$ in.^{3}, and $I = 697.068$ in.^{4}?

A. 10,120 ft.-lbs.

B. 11,153 ft.-lbs.

C. 16,164 ft.-lbs.

D. 19,397 ft.-lbs.

245. Long span steel joists with a clear span of 80 feet are used to support a roof. The joists are six feet apart, the dead load is 20 pounds per square foot (including the joist weight), the live load is 30 pounds per square foot, and the live load deflection is limited to 1/360 of the span. What is the lightest joist that can be used? Use the table on page 64.

A. 40LH12

B. 40LH13

C. 44LH13

D. 48LH12

246. Which of the following statements concerning strength design of reinforced concrete is *INCORRECT*?

A. The amount of reinforcing steel used assures that yielding of the steel will occur before failure of the concrete.

B. The reinforcing steel is generally assumed to resist all the tensile stresses, while the concrete is assumed to resist all the compressive stresses.

C. The amount of reinforcing steel used must be at least equal to that which would produce a balanced design.

D. The ultimate load factors are greater for live load than for dead load.

247. A steel beam spans 30 feet and supports a load of 1,800 pounds per foot including the weight of the beam. What is the lightest steel section that can support the load? Assume A36 steel and full lateral support. Use the table on page 65.

   A. W14 × 48

   B. W14 × 68

   C. W21 × 44

   D. W24 × 55

248. What is a stub girder system?

   A. A steel beam-and-girder system in which floor beams pass over the main girders and stub girders are welded over the main girders along the same axis

   B. Short lengths of steel girders shop welded to steel columns to simplify field erection

   C. Wood or steel girders designed and detailed for simulated continuity

   D. Steel girders used at the base of a steel column to spread the column load

249. Select the correct statements.

   I.   The modulus of elasticity of concrete varies with the strength and weight of the concrete.

   II. The deflection of a reinforced concrete beam increases over time, even if the load is not increased.

   III. Adding compressive reinforcement has no effect on the deflection of a reinforced concrete beam.

   A. I and II

   B. II and III

   C. I and III

   D. I, II, and III

250. What is the internal force in member BD of the truss shown?

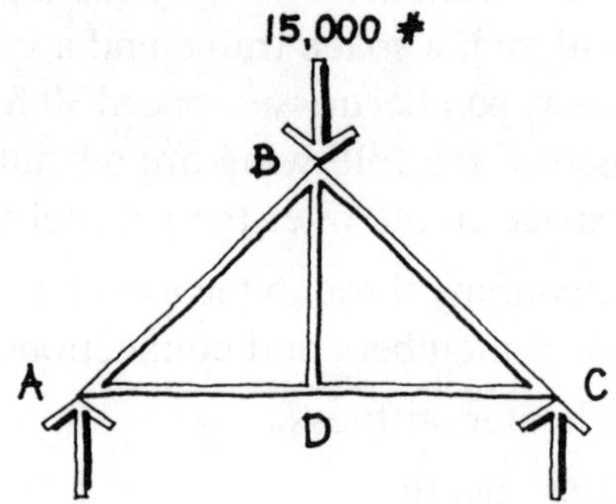

   A. Zero

   B. 5 kips

   C. 7.5 kips

   D. 15 kips

251. Select the correct statement about composite design.

   A. Connectors welded to the steel beam and embedded in the concrete resist flexural stresses.

   B. Composite design is most efficient with light loads and short spans.

   C. The concrete slab resists compressive bending stresses.

   D. Because a composite beam is stiffer than a non-composite beam, it is usually unnecessary to check deflections.

252. Why are stirrups used in reinforced concrete construction?

   A. To provide compressive reinforcement

   B. To provide web reinforcement where the concrete is overstressed in shear

   C. To anchor the tensile reinforcement

   D. To prevent lateral buckling of compressive reinforcement

253. What is the maximum load P that can be transferred by the splice detail shown above? The members are Douglas Fir-Larch. Use the table on page 67.

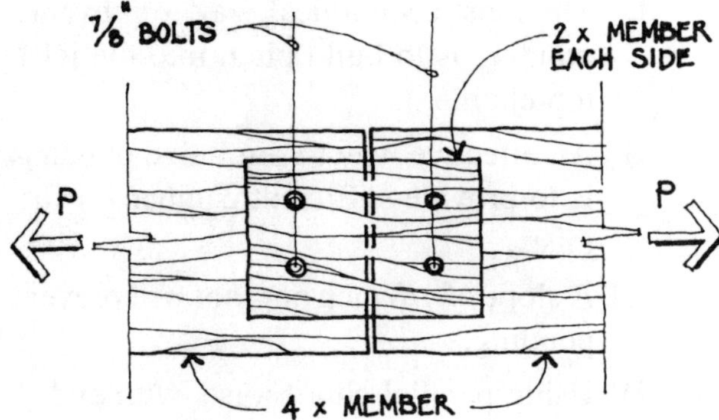

A. 2,520#

B. 2,780#

C. 5,600#

D. 7,160#

254. What property is practically constant for all structural steels?

A. Yield point

B. Ultimate strength

C. Weldability

D. Modulus of elasticity

255. What is a lamella roof?

A. A series of parallel arches, skewed with respect to the axes of a building, which intersect another series of skewed arches

B. A two-way truss system in which the trusses are inclined, rather than in a vertical plane

C. A series of radial cables stabilized by another series of cables

D. A system consisting of two intersecting arches placed along the diagonals of a building

256. Which of the following statements are correct concerning an air-supported fabric structure which encloses an occupied space?

  I. Fans are required to pressurize the space.

  II. People and equipment must pass through airlocks or special doors to get into and out of the facility.

  III. Long spans require steel cable reinforcement.

  IV. Loss of pressure can cause the roof to deflate and possibly fail.

A. I and III

B. II and IV

C. I, II, and IV

D. I, II, III, and IV

257. Two angles placed as shown are to be used for the top chord of a truss whose panel points are 10 feet apart. To determine the required size of the angles, the value of the radius of gyration r should be with respect to which axis?

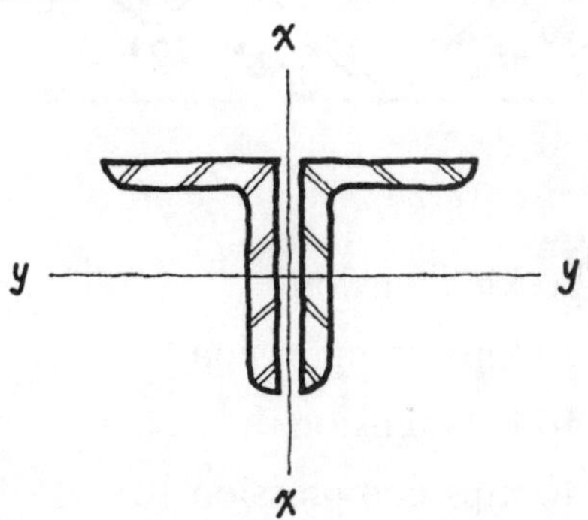

A. x-x

B. y-y

C. x-x or y-y, whichever results in the lower value of r

D. x-x or y-y, whichever results in the higher value of r

258. Which of the diagrams below represents the flexural stresses in a reinforced concrete beam at failure?

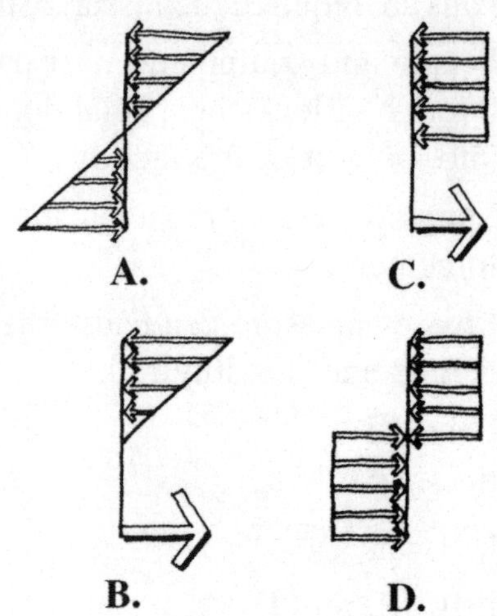

259. What is the internal force in member a of the cantilever truss shown?

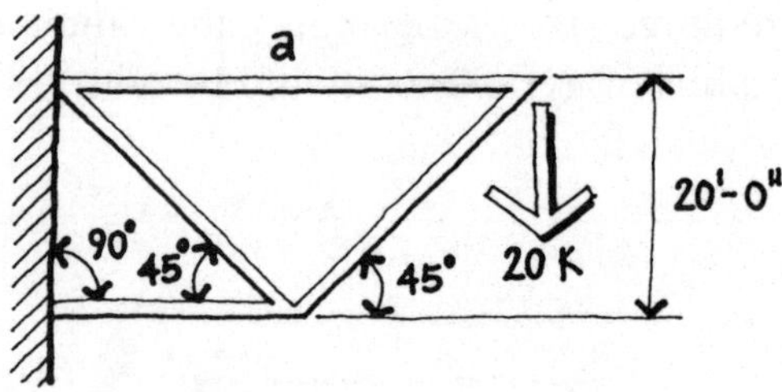

A. 20 kips tension

B. 20 kips compression

C. 40 kips tension

D. 40 kips compression

260. Which type of arch is statically determinate when subject to a uniformly distributed load across the entire span?

A. A two-hinged arch

B. A three-hinged arch

C. A fixed arch

D. Any arch

261. The ponding of water on a flat roof supported by long span open web joists can be a problem. In this regard, which of the following statements are correct?

I.   The most economical way to prevent ponding is to build pitch into the joist top chords.

II. An effective way to minimize ponding is to provide sufficient camber for the joists.

III. A slope of 1/8 inch per foot will prevent ponding.

IV. Using parallel chord joists with end supports at different elevations is an economical way to provide slope.

A. I and II

B. II and III

C. II, III and IV

D. I and III

262. What is the horizontal thrust at each end of the three-hinged arch shown below?

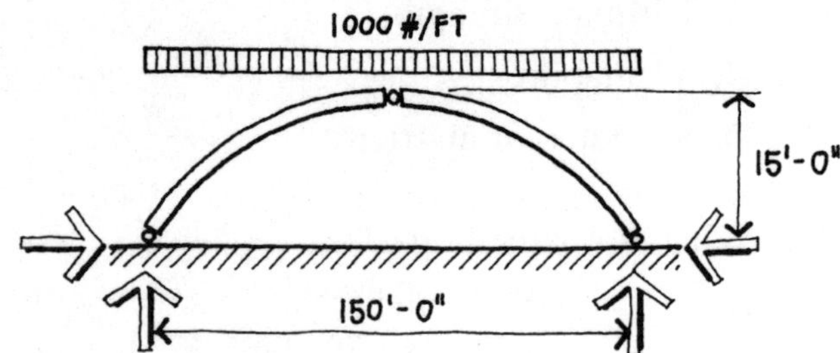

A. Not determinable from the information given

B. 75 kips

C. 187.5 kips

D. 375 kips

263. In glued laminated construction, it is often necessary to join pieces of lumber end to end to produce laminations of sufficient length. Which of the following joints are acceptable for this purpose?

I.

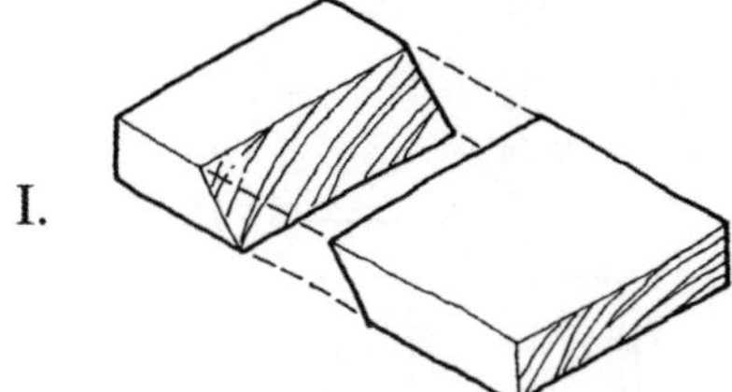

PLAIN SCARF JOINT

II.

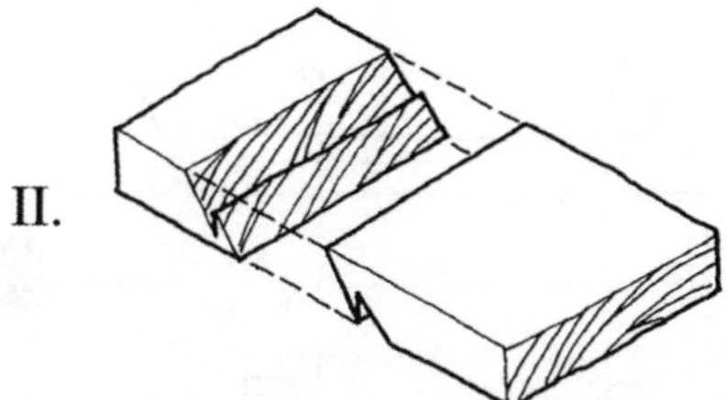

HOOKED SCARF JOINT

III.

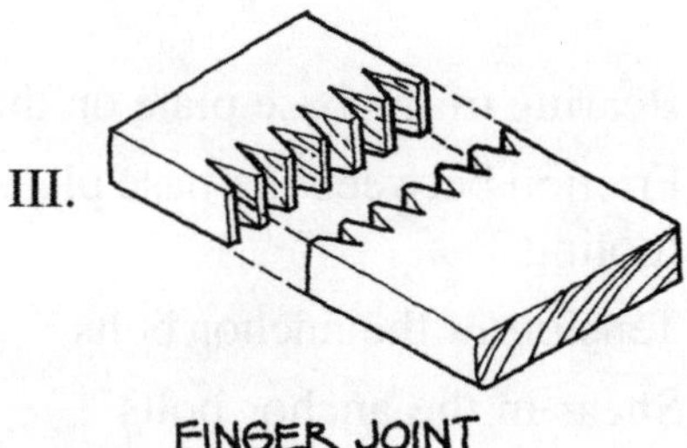

FINGER JOINT

- **A.** I and II
- **B.** I and III
- **C.** II and III
- **D.** I, II, and III

264. Which of the following statements is *INCORRECT*? The flat slab floor system

- **A.** spans simultaneously in two directions.
- **B.** is economical in reinforcing steel.
- **C.** is appropriate for heavy live loads.
- **D.** generally has no beams or girders.

265. What is a composite deck?

- **A.** Steel decking and a concrete slab, which act together to span between beams
- **B.** Decking made of two different grades of steel
- **C.** A steel beam and a formed concrete slab, which act together as a single flexural member
- **D.** A steel beam with steel decking and a concrete slab over, which act together as a single flexural member

266. The nominal diameter and yield strength of the reinforcing bar shown below are as follows:

- **A.** Diameter = 5/8 inch, yield strength = 60 ks
- **B.** Diameter = 1 inch, yield strength = 60 ksi
- **C.** Diameter = 0.60 inch, yield strength = 100 ksi
- **D.** Diameter = 1-1/4 inch, yield strength = 60 ksi

267. I-shaped wood joists consist of lumber flanges and a strand board web inserted and glued into a groove in each flange. What type of stress does the glue resist?

- **A.** Flexural tension or compression
- **B.** Axial tension or compression
- **C.** Bending moment
- **D.** Horizontal shear

268. A hole or notch for a pipe must be provided in a reinforced concrete beam. Which of the diagrams below shows the hole or notch that will *LEAST* affect the beam's load-carrying capacity?

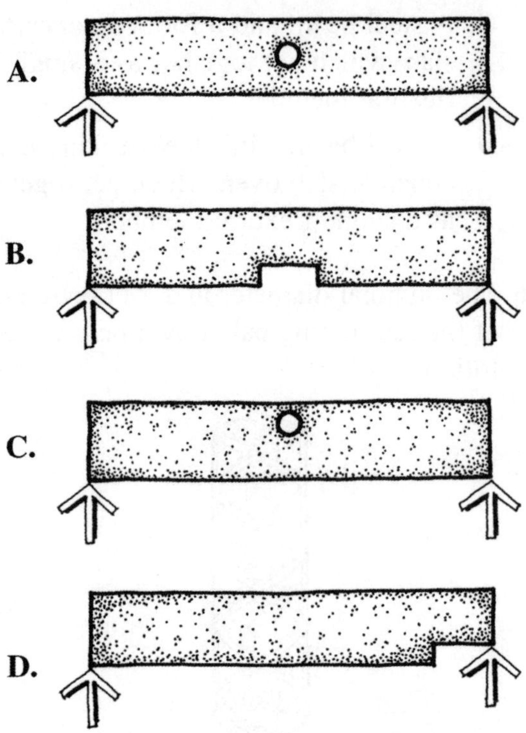

269. Which of the following statements about prestressed concrete construction is *NOT* true?

A. Precast, prestressed members usually require end anchorages.

B. Pretensioned members require no end anchorages.

C. Prestressing a beam results in a smaller section.

D. Posttensioned members usually require end anchorages.

270. A steel column supports a vertical load V and resists a horizontal shear H as shown. How is the horizontal shear transferred from the column to the footing?

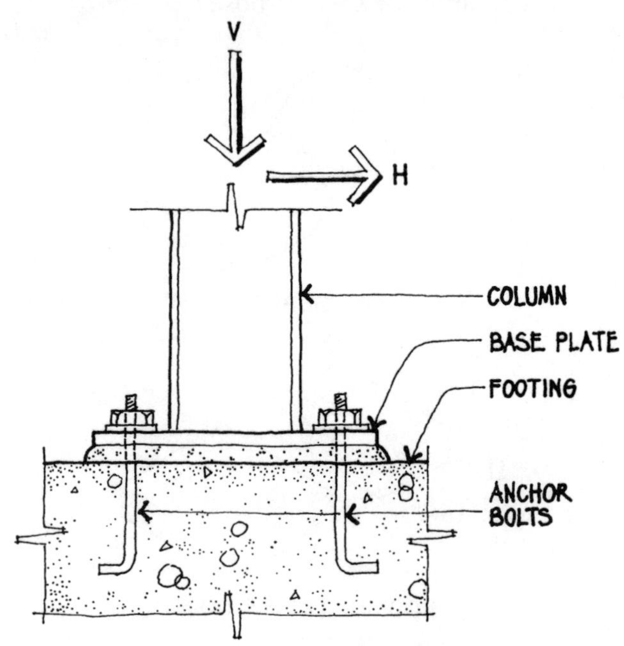

A. Bearing of the base plate on the footing

B. Friction between the base plate and the footing

C. Tension in the anchor bolts

D. Shear in the anchor bolts

271. Where is a wood beam most likely to fail in shear?

I.   At midspan

II. Near the supports

III. Close to mid-depth of the beam

IV. Close to the top of the beam

A. I and IV

B. II and III

C. I and III

D. II and IV

272. A cantilever beam is loaded as shown. What is the magnitude of the reaction V and the moment M at the fixed end, neglecting the weight of the beam?

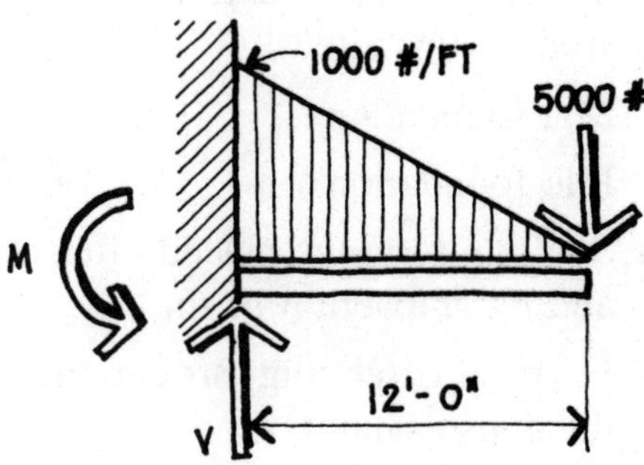

A. V = 11,000#; M = 84,000 ft.-lbs.

B. V = 11,000#; M = 96,000 ft.-lbs.

C. V = 17,000#; M = 108,000 ft.-lbs.

D. V = 17,000#; M = 132,000 ft.-lbs.

273. The load-carrying capacity of a wood column is determined by all of the following, *EXCEPT*

A. the species and grade of lumber.

B. the dimensions of the column section.

C. the unbraced height of the column.

D. the applied load.

274. The strength of a complete penetration groove weld in tension or compression is

A. based on the shear strength through the throat of the weld.

B. equal to that of a fillet weld.

C. the same as that of the connected material.

D. less than that of the connected material.

275. Which of the following statements concerning the maximum size of coarse aggregate in a concrete mixture is *INCORRECT?*

A. The maximum size of coarse aggregate is limited by the narrowest dimension between forms.

B. The maximum size of coarse aggregate is less than the minimum clear spacing between reinforcing bars.

C. The smaller the maximum size of coarse aggregate, the greater the amount of water required.

D. For economy, the maximum size of coarse aggregate should be as small as possible.

276. A stub girder system is analogous to

A. a triangulated truss.

B. a rigid frame.

C. a Vierendeel truss.

D. an open web joist.

277. A column of ASTM A36 steel is 16 feet long and supports a load of 200 kips. What is the most economical W10 column section that can support the load? Use the chart on page 68.

A. W10 × 39

B. W10 × 45

C. W10 × 49

D. W10 × 54

278. Two different framing schemes are being considered, as shown below. Scheme A consists of three simple beams of equal span. In scheme B, the middle beam is hung from the outer beams. What are the advantages of scheme B?

I.   Reduced positive moment in the end spans

II. Reduced positive moment in the center span

III. Reduced column loads

A. I only

B. I and II

C. II only

D. I, II, and III

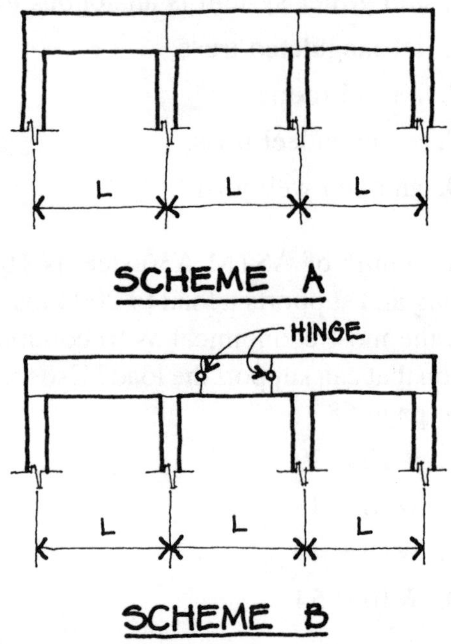

279. The soil boring log for a level building site shows that the top six feet of subsurface material consists of a loose fill, below which is a thick layer of dense sand. Which of the following foundation systems could be used on this site?

I.   Mat foundation

II. Pile foundation

III. Shallow footings placed after removal and recompaction of the fill

IV. Footings extending through the fill into the dense sand

A. I only

B. II and III

C. I, II, and IV

D. I, II, III, and IV

280. The load applied to a retaining wall by the retained earth and the resulting soil pressure under its footing are most closely approximated by which of the following diagrams?

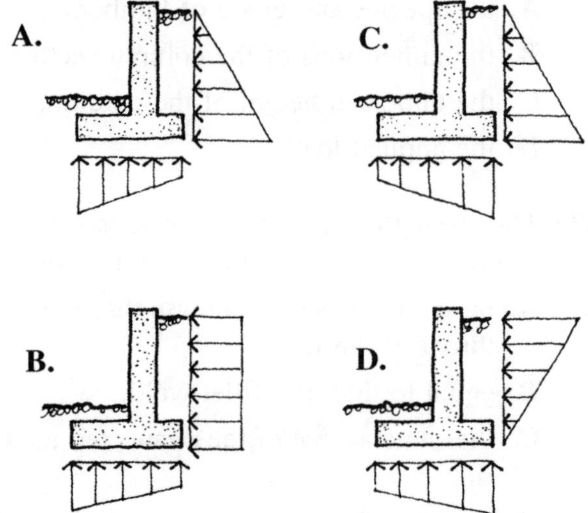

281. A steel column in a moment-resisting steel frame is to be constructed as shown below. To determine the value of K for the column, the top of the column is assumed to be

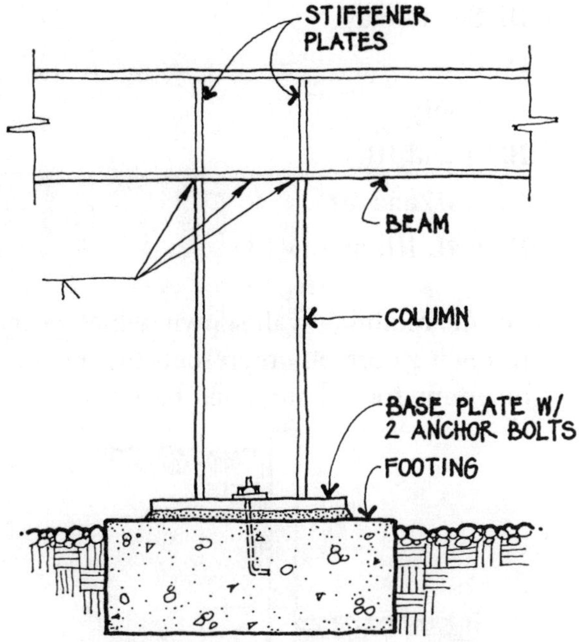

A. fixed against rotation and free to translate.

B. free to rotate and fixed against translation.

C. free to rotate and free to translate.

D. fixed against rotation and fixed against translation.

282. Which of the following types of soil is most susceptible to volumetric change caused by change of moisture content?

A. Sand

B. Gravel

C. Rock

D. Clay

283. Complete the following statement. A counterfort wall is

A. a retaining wall that depends entirely on its own weight to resist the pressure of the retained earth.

B. a retaining wall formed of stacked elements that form cells that are filled with soil.

C. a retaining wall consisting of a vertical stem supported at intervals on the back side by triangular buttresses connected to the base.

D. a retaining wall consisting of a vertical stem and a base that are reinforced to resist the moments caused by the pressure of the retained earth.

284. For the retaining wall shown, which faces are in tension (noted as T) and which are in compression (noted as C)?

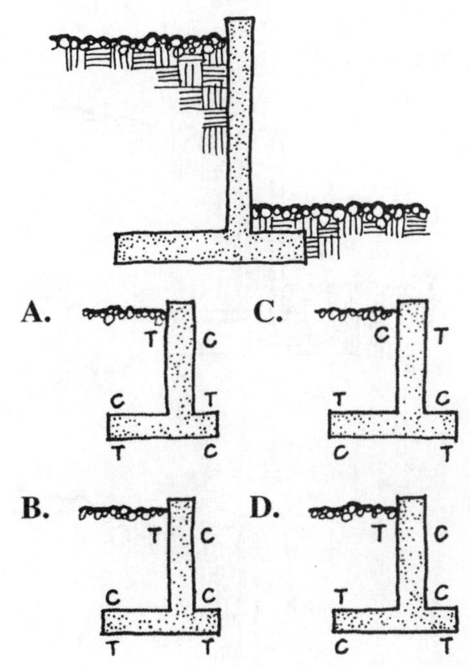

285. Which of the following retaining walls has the reinforcing steel *INCORRECTLY* placed?

A.

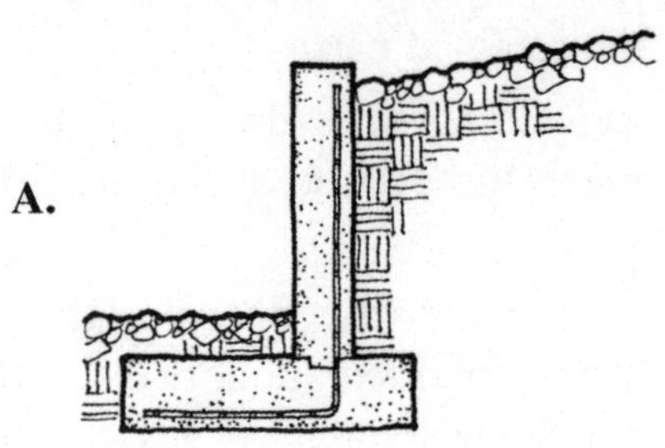

B.

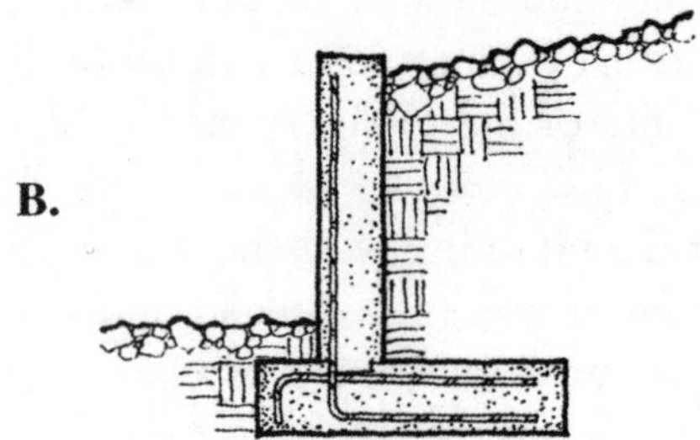

C.

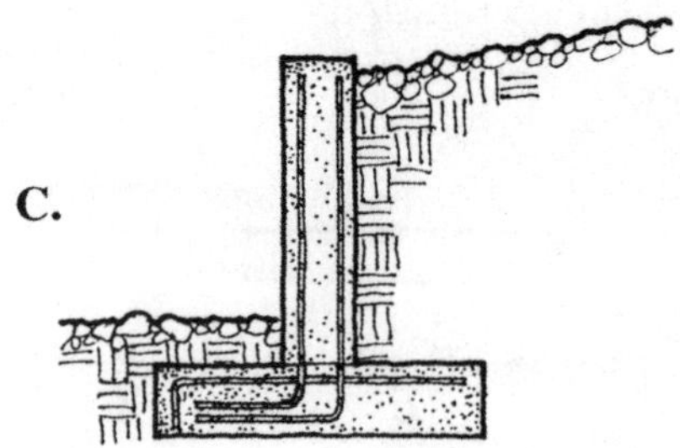

D.

286. Which of the following soils may be used to support building foundations?

   I. Sandy gravel

   II. Silty sand

   III. Sandy clay

   IV. Peat

  A. I only

  B. II and III

  C. I, II, and III

  D. I, II, III, and IV

287. For the retaining wall shown, which of the following correctly represents the pressure diagrams for the stem and base?

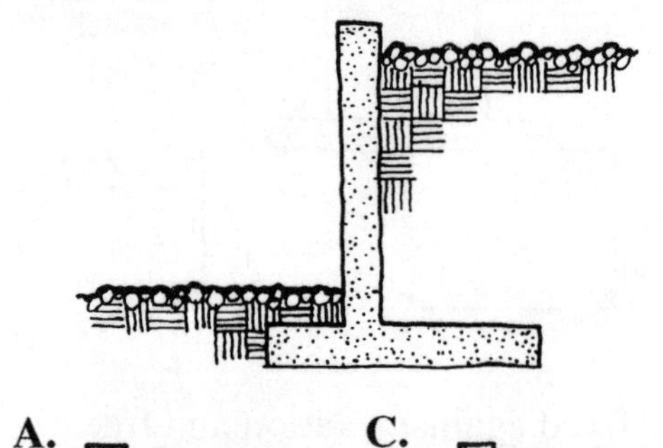

A.   C.

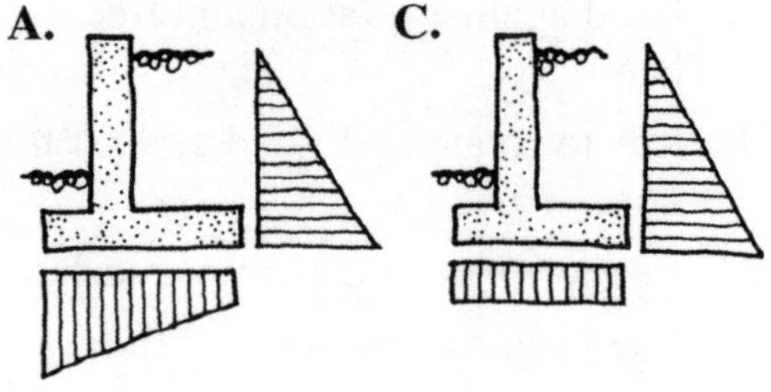

B.   D.

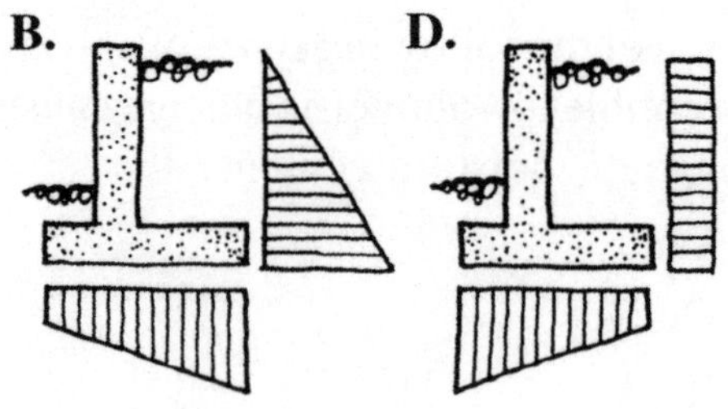

288. A steel column supports a dead load of 120 kips and a live load of 150 kips. The allowable soil bearing value is 4,000 pounds per square foot. What is the smallest pad footing that may be used?

A. 5'-6" × 5'-6"

B. 6'-2" × 6'-2"

C. 6'-9" × 6'-9"

D. 8'-3" × 8'-3"

289. What is the total lateral force exerted by the earth against the retaining wall shown, per lineal foot of wall? Assume the pressure of the retained earth to be equivalent to a fluid weighing 30 pounds per cubic foot.

A. 150#

B. 300#

C. 1,500#

D. 3,000#

290. All of the following may increase the sliding resistance of a retaining wall, *EXCEPT*

A. adding an integral key to the footing.

B. making the footing wider.

C. making the footing deeper.

D. increasing the amount of reinforcing steel in the footing.

291. What is the purpose of the footing shown?

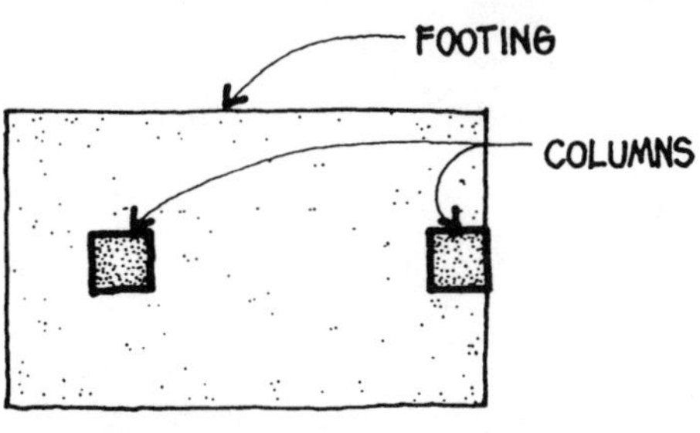

A. To resist differential settlement

B. To fix the column bases against rotation

C. To distribute the column loads over a large area when soil conditions are poor

D. To support two columns where one of the columns is too close to the property line to have a symmetrical footing

292. The soil boring log for a building site shows that the upper 15 feet of subsurface material is a loose fill, below which is a thick layer of dense sand. Which of the following foundation systems might be appropriate on this site?

I.   Footings placed after the fill is removed and recompacted

II. Footings extending through the fill into the dense sand

III. Belled caissons bearing on the dense sand

IV. Piles extending through the fill into the dense sand

A. I and II

B. III and IV

C. II, III, and IV

D. I, II, III, and IV

The Architect Registration Examination (ARE) is closed book; no reference material is permitted for any part of the exam. Any reference material required for the structural test will be provided when you take the exam.

We suggest that, prior to the exam, you obtain some of the reference books that may be excerpted, in order to become familiar with them and thereby save valuable time at the exam.

The reference books that have often had sections reproduced for reference during the test include the *AISC Manual*, the *Uniform Building Code,* and the *Standard Specifications for Steel Joists.*

Many architectural and engineering offices have these references in their library, which you may be able to borrow during your exam preparation.

Once you have the reference books, you should try to become generally familiar with the scope and format of each book, how its charts and tables are organized, and so forth, rather than trying to memorize anything.

On the following pages, we have reproduced the charts and tables needed to answer several of the questions in this book.

BOLTS, THREADED PARTS AND RIVETS
Shear Allowable Load in Kips

TABLE 1-D. SHEAR

	ASTM Desig-nation	Conn-ection Type[a]	Hole Type[b]	F_u ksi	Load-ing[c]	Nominal Diameter d, in.							
						5/8	3/4	7/8	1	1 1/8	1 1/4	1 3/8	1 1/2
						Area (Based on Nominal Diameter) in.[2]							
						.3068	.4418	.6013	.7854	.9940	1.227	1.485	1.767
Bolts	A307	—	STD NSL	10.0	S	3.1	4.4	6.0	7.9	9.9	12.3	14.8	17.7
					D	6.1	8.8	12.0	15.7	19.9	24.5	29.7	35.3
	A325	SC[a] Class A	STD	17.0	S	5.22	7.51	10.2	13.4	16.9	20.9	25.2	30.0
					D	10.4	15.0	20.4	26.7	33.8	41.7	50.5	60.1
			OVS, SSL	15.0	S	4.60	6.63	9.02	11.8	14.9	18.4	22.3	26.5
					D	9.20	13.3	18.0	23.6	29.8	36.8	44.6	53.0
			LSL	12.0	S	3.68	5.30	7.22	9.42	11.9	14.7	17.8	21.2
					D	7.36	10.6	14.4	18.8	23.9	29.4	35.6	42.4
		N	STD, NSL	21.0	S	6.4	9.3	12.6	16.5	20.9	25.8	31.2	37.1
					D	12.9	18.6	25.3	33.0	41.7	51.5	62.4	74.2
		X	STD, NSL	30.0	S	9.2	13.3	18.0	23.6	29.8	36.8	44.5	53.0
					D	18.4	26.5	36.1	47.1	59.6	73.6	89.1	106.0
	A490	SC[a] Class A	STD	21.0	S	6.44	9.28	12.6	16.5	20.9	25.8	31.2	37.1
					D	12.9	18.6	25.3	33.0	41.7	51.5	62.4	74.2
			OVS, SSL	18.0	S	5.52	7.95	10.8	14.1	17.9	22.1	26.7	31.8
					D	11.0	15.9	21.6	28.3	35.8	44.2	53.5	63.6
			LSL	15.0	S	4.60	6.63	9.02	11.8	14.9	18.4	22.3	26.5
					D	9.20	13.3	18.0	23.6	29.8	36.8	44.6	53.0
		N	STD, NSL	28.0	S	8.6	12.4	16.8	22.0	27.8	34.4	41.6	49.5
					D	17.2	24.7	33.7	44.0	55.7	68.7	83.2	99.0
		X	STD, NSL	40.0	S	12.3	17.7	24.1	31.4	39.8	49.1	59.4	70.7
					D	24.5	35.3	48.1	62.8	79.5	98.2	119.0	141.0
Rivets	A502-1	—	STD	17.5	S	5.4	7.7	10.5	13.7	17.4	21.5	26.0	30.9
					D	10.7	15.5	21.0	27.5	34.8	42.9	52.0	61.8
	A502-2 A502-3	—	STD	22.0	S	6.7	9.7	13.2	17.3	21.9	27.0	32.7	38.9
					D	13.5	19.4	26.5	34.6	43.7	54.0	65.3	77.7
Threaded Parts	A36 (F_u=58 ksi)	N	STD	9.9	S	3.0	4.4	6.0	7.8	9.8	12.1	14.7	17.5
					D	6.1	8.7	11.9	15.6	19.7	24.3	29.4	35.0
		X	STD	12.8	S	3.9	5.7	7.7	10.1	12.7	15.7	19.0	22.6
					D	7.9	11.3	15.4	20.1	25.4	31.4	38.0	45.2
	A572, Gr. 50 (F_u=65 ksi)	N	STD	11.1	S	3.4	4.9	6.7	8.7	11.0	13.6	16.5	19.6
					D	6.8	9.8	13.3	17.4	22.1	27.2	33.0	39.2
		X	STD	14.3	S	4.4	6.3	8.6	11.2	14.2	17.5	21.2	25.3
					D	8.8	12.6	17.2	22.5	28.4	35.1	42.5	50.5
	A588 (F_u=70 ksi)	N	STD	11.9	S	3.7	5.3	7.2	9.3	11.8	14.6	17.7	21.0
					D	7.3	10.5	14.3	18.7	23.7	29.2	35.3	42.1
		X	STD	15.4	S	4.7	6.8	9.3	12.1	15.3	18.9	22.9	27.2
					D	9.4	13.6	18.5	24.2	30.6	37.8	45.7	54.4

[a]SC = Slip critical connection.
 N: Bearing-type connection with threads *included* in shear plane.
 X: Bearing-type connection with threads *excluded* from shear plane.
[b]STD: Standard round holes ($d + 1/16$ in.)   OVS: Oversize round holes
 LSL: Long-slotted holes   SSL: Short-slotted holes
 NSL: Long-or short-slotted hole normal to load direction
      (required in bearing-type connection).
[c]S: Single shear   D: Double shear.
For threaded parts of materials not listed, use $F_v = 0.17F_u$ when threads are included in a shear plane, and $F_v = 0.22F_u$ when threads are excluded from a shear plane.
To fully pretension bolts 1 1/8-in. dia. and greater, special impact wrenches may be required.
When bearing-type connections used to splice tension members have a fastener pattern whose length, measured parallel to the line of force, exceeds 50 in., tabulated values shall be reduced by 20%. See AISC ASD Commentary Sect. J3.4.

BOLTS AND THREADED PARTS
Bearing Allowable Loads in Kips

TABLE 1-E. BEARING											
Slip-Critical and Bearing-Type Connections											
Mate-rial Thick-ness	F_u = 58 ksi Bolt dia.			F_u = 65 ksi Bolt dia.			F_u = 70 ksi Bolt dia.			F_u = 100 ksi Bolt dia.		
	3/4	7/8	1	3/4	7/8	1	3/4	7/8	1	3/4	7/8	1
1/8	6.5	7.6	8.7	7.3	8.5	9.8	7.9	9.2	10.5	11.3	13.1	15.0
3/16	9.8	11.4	13.1	11.0	12.8	14.6	11.8	13.8	15.8	16.9	19.7	22.5
1/4	13.1	15.2	17.4	14.6	17.1	19.5	15.8	18.4	21.0	22.5	26.3	30.0
5/16	16.3	19.0	21.8	18.3	21.3	24.4	19.7	23.0	26.3	28.1	32.8	37.5
3/8	19.6	22.8	26.1	21.9	25.6	29.3	23.6	27.6	31.5	33.8	39.4	45.0
7/16	22.8	26.6	30.5	25.6	29.9	34.1	27.6	32.2	36.8		45.9	52.5
1/2	26.1	30.5	34.8	29.3	34.1	39.0	31.5	36.8	42.0			60.0
9/16	29.4	34.3	39.2	32.9	38.4	43.9		41.3	47.3			
5/8	32.6	38.1	43.5		42.7	48.8		45.9	52.5			
11/16		41.9	47.9		46.9	53.6			57.8			
3/4		45.7	52.2			58.5						
13/16			56.6									
7/8			60.9									
15/16												
1	52.2	60.9	69.6	58.5	68.3	78.0	63.0	73.5	84.0	90.0	105.0	120.0

Notes:

This table is applicable to all mechanical fasteners in both slip-critical and bearing-type connections utilizing standard holes. Standard holes shall have a diameter nominally 1/16-in. larger than the nominal bolt diameter (d + 1/16 in.).

Tabulated bearing values are based on F_p = 1.2 F_u.

F_u = specified minimum tensile strength of the connected part.

In connections transmitting axial force whose length between extreme fasteners measured parallel to the line of force exceeds 50 in., tabulated values shall be reduced 20%.

Connections using high-strength bolts in slotted holes with the load applied in a direction other than approximately normal (between 80 and 100 degrees) to the axis of the hole and connections with bolts in oversize holes shall be designed for resistance against slip at working load in accordance with AISC ASD Specification Sect. J3.8.

Tabulated values apply when the distance l parallel to the line of force from the center of the bolt to the edge of the connected part is not less than 1½ d and the distance from the center of a bolt to the center of an adjacent bolt is not less than 3d. See AISC ASD Commentary J3.8.

Under certain conditions, values greater than the tabulated values may be justified under Specification Sect. J3.7.

Values are limited to the double-shear bearing capacity of A490-X bolts.

Values for decimal thicknesses may be obtained by multiplying the decimal value of the unlisted thickness by the value given for a 1-in. thickness.

STANDARD LOAD TABLE/LONGSPAN STEEL JOISTS, LH-SERIES

Based on a Maximum Allowable Tensile Stress of 30 ksi

Joist Designation	Approx. Wt. in Lbs. per Linear Ft. (Joists Only)	Depth in Inches	SAFE LOAD* in Lbs. Between 47-59	60-64	65	66	67	68	69	70	71	72	73	74	75	76	77	78	79	80
									CLEAR SPAN IN FEET											
40LH08	16	40	16600	16600	254 150	247 144	241 138	234 132	228 127	222 122	217 117	211 112	206 108	201 104	196 100	192 97	187 93	183 90	178 86	174 83
40LH09	21	40	21800	21800	332 196	323 188	315 180	306 173	298 166	291 160	283 153	276 147	269 141	263 136	256 131	250 126	244 122	239 118	233 113	228 109
40LH10	21	40	24000	24000	367 216	357 207	347 198	338 190	329 183	321 176	313 169	305 162	297 156	290 150	283 144	276 139	269 134	262 129	255 124	249 119
40LH11	22	40	26200	26200	399 234	388 224	378 215	368 207	358 198	349 190	340 183	332 176	323 169	315 163	308 157	300 151	293 145	286 140	279 135	273 130
40LH12	25	40	31900	31900	486 285	472 273	459 261	447 251	435 241	424 231	413 222	402 213	392 205	382 197	373 189	364 182	355 176	346 169	338 163	330 157
40LH13	30	40	37600	37600	573 334	557 320	542 307	528 295	514 283	500 271	487 260	475 250	463 241	451 231	440 223	429 214	419 207	409 199	399 192	390 185
40LH14	35	40	43000	43000	656 383	638 367	620 351	603 336	587 323	571 309	556 297	542 285	528 273	515 263	502 252	490 243	478 233	466 225	455 216	444 209
40LH15	36	40	48100	48100	734 427	712 408	691 390	671 373	652 357	633 342	616 328	599 315	583 302	567 290	552 279	538 268	524 258	511 248	498 239	486 230
40LH16	42	40	53000	53000	808 469	796 455	784 441	772 428	761 416	751 404	730 387	710 371	691 356	673 342	655 329	638 316	622 304	606 292	591 282	576 271

Joist Designation	Approx. Wt. in Lbs. per Linear Ft. (Joists Only)	Depth in Inches	SAFE LOAD* in Lbs. Between 52-59	60-72	73	74	75	76	77	78	79	80	81	82	83	84	85	86	87	88
44LH09	19	44	20000	20000	272 158	265 152	259 146	253 141	247 136	242 131	236 127	231 122	226 118	221 114	216 110	211 106	207 103	202 99	198 96	194 93
44LH10	21	44	22100	22100	300 174	293 168	286 162	279 155	272 150	266 144	260 139	254 134	249 130	243 125	238 121	233 117	228 113	223 110	218 106	214 103
44LH11	22	44	23900	23900	325 188	317 181	310 175	302 168	295 162	289 157	282 151	276 146	269 140	264 136	258 131	252 127	247 123	242 119	236 115	232 111
44LH12	25	44	29600	29600	402 232	393 224	383 215	374 207	365 200	356 192	347 185	339 179	331 172	323 166	315 160	308 155	300 149	293 144	287 139	280 134
44LH13	30	44	35100	35100	477 275	466 265	454 254	444 246	433 236	423 228	413 220	404 212	395 205	386 198	377 191	369 185	361 179	353 173	346 167	338 161
44LH14	31	44	40400	40400	549 315	534 302	520 291	506 279	493 268	481 259	469 249	457 240	446 231	436 223	425 215	415 207	406 200	396 193	387 187	379 181
44LH15	36	44	47000	47000	639 366	623 352	608 339	593 326	579 314	565 303	551 292	537 281	524 271	512 261	500 252	488 243	476 234	466 227	455 219	445 211
44LH16	42	44	54200	54200	737 421	719 405	701 390	684 375	668 362	652 348	637 336	622 324	608 313	594 302	580 291	568 282	555 272	543 263	531 255	520 246
44LH17	47	44	58200	58200	790 450	780 438	769 426	759 415	750 405	732 390	715 376	699 363	683 351	667 338	652 327	638 316	624 305	610 295	597 285	584 276

Joist Designation	Approx. Wt. in Lbs. per Linear Ft. (Joists Only)	Depth in Inches	SAFE LOAD* in Lbs. Between 56-59	60-80	81	82	83	84	85	86	87	88	89	90	91	92	93	94	95	96
48LH10	21	48	20000	20000	246 141	241 136	236 132	231 127	226 123	221 119	217 116	212 112	208 108	204 105	200 102	196 99	192 96	188 93	185 90	181 87
48LH11	22	48	21700	21700	266 152	260 147	255 142	249 137	244 133	239 129	234 125	229 120	225 117	220 113	216 110	212 106	208 103	204 100	200 97	196 94
48LH12	25	48	27400	27400	336 191	329 185	322 179	315 173	308 167	301 161	295 156	289 151	283 147	277 142	272 138	266 133	261 129	256 126	251 122	246 118
48LH13	29	48	32800	32800	402 228	393 221	384 213	376 206	368 199	360 193	353 187	345 180	338 175	332 170	325 164	318 159	312 154	306 150	300 145	294 141
48LH14	32	48	38700	38700	475 269	464 260	454 251	444 243	434 234	425 227	416 220	407 212	399 206	390 199	383 193	375 187	367 181	360 176	353 171	346 165
48LH15	36	48	44500	44500	545 308	533 298	521 287	510 278	499 269	488 260	478 252	468 244	458 236	448 228	439 221	430 214	422 208	413 201	405 195	397 189
48LH16	42	48	51300	51300	629 355	615 343	601 331	588 320	576 310	563 299	551 289	540 280	528 271	518 263	507 255	497 247	487 239	477 232	468 225	459 218
48LH17	47	48	57600	57600	706 397	690 383	675 371	660 358	646 346	632 335	619 324	606 314	593 304	581 294	569 285	558 276	547 268	536 260	525 252	515 245

ALLOWABLE STRESS DESIGN SELECTION TABLE
For shapes used as beams S_x

$F_y = 50$ ksi			S_x	Shape	Depth d	F'_y	$F_y = 36$ ksi		
L_c	L_u	M_R					L_c	L_u	M_R
Ft	Ft	Kip-ft	In.³		In.	Ksi	Ft	Ft	Kip-ft
8.1	**8.6**	**484**	**176**	**W 24× 76**	**23⅞**	—	**9.5**	**11.8**	**348**
9.3	20.2	481	175	W 16×100	17	—	11.0	28.1	347
13.1	29.2	476	173	W 14×109	14⅜	58.6	15.4	40.6	343
7.5	10.9	470	171	W 21× 83	21⅜	—	8.8	15.1	339
9.9	15.5	457	166	W 18× 86	18⅜	—	11.7	21.5	329
13.0	26.7	432	157	W 14× 99	14⅛	48.5	15.4	37.0	311
9.3	18.0	426	155	W 16× 89	16¾	—	10.9	25.0	307
7.4	**8.5**	**424**	**154**	**W 24× 68**	**23¾**	—	**9.5**	**10.2**	**305**
7.4	9.6	415	151	W 21× 73	21¼	—	8.8	13.4	299
9.9	13.7	402	146	W 18× 76	18¼	64.2	11.6	19.1	289
13.0	24.5	385	143	W 14× 90	14	40.4	15.3	34.0	283
7.4	**8.9**	**385**	**140**	**W 21× 68**	**21⅛**	—	**8.7**	**12.4**	**277**
9.2	15.8	369	134	W 16× 77	16½	—	10.9	21.9	265
5.8	**6.4**	**360**	**131**	**W 24× 62**	**23¾**	—	**7.4**	**8.1**	**259**
7.4	**8.1**	**349**	**127**	**W 21× 62**	**21**	—	**8.7**	**11.2**	**251**
6.8	11.1	349	127	W 18× 71	18½	—	8.1	15.5	251
9.1	20.2	338	123	W 14× 82	14¼	—	10.7	28.1	244
10.9	26.0	325	118	W 12× 87	12½	—	12.8	36.2	234
6.8	10.4	322	117	W 18× 65	18⅜	—	8.0	14.4	232
9.2	13.9	322	117	W 16× 67	16⅜	—	10.8	19.3	232
5.0	**6.3**	**314**	**114**	**W 24× 55**	**23⅜**	—	**7.0**	**7.5**	**226**
9.0	18.6	308	112	W 14× 74	14⅛	—	10.6	25.9	222
5.9	6.7	305	111	W 21× 57	21	—	6.9	9.4	220
6.8	9.6	297	108	W 18× 60	18¼	—	8.0	13.3	214
10.8	24.0	294	107	W 12× 79	12⅜	62.6	12.8	33.3	212
9.0	17.2	283	103	W 14× 68	14	—	10.6	23.9	204
6.7	**8.7**	**270**	**98.3**	**W 18× 55**	**18⅛**	—	**7.9**	**12.1**	**195**
10.8	21.9	268	97.4	W 12× 72	12¼	52.3	12.7	30.5	193
5.6	**6.0**	**260**	**94.5**	**W 21× 50**	**20⅞**	—	**6.9**	**7.8**	**187**
6.4	10.3	254	92.2	W 16× 57	16⅜	—	7.5	14.3	183
9.0	15.5	254	92.2	W 14× 61	13⅞	—	10.6	21.5	183
6.7	**7.9**	**244**	**88.9**	**W 18× 50**	**18**	—	**7.9**	**11.0**	**176**
10.7	20.0	238	87.9	W 12× 65	12⅛	43.0	12.7	27.7	174
4.7	**5.9**	**224**	**81.6**	**W 21× 44**	**20⅝**	—	**6.6**	**7.0**	**162**
6.3	9.1	223	81.0	W 16× 50	16¼	—	7.5	12.7	160
5.4	6.8	217	78.8	W 18× 46	18	—	6.4	9.4	156
9.0	17.5	215	78.0	W 12× 58	12¼	—	10.6	24.4	154
7.2	12.7	214	77.8	W 14× 53	13⅞	—	8.5	17.7	154
6.3	8.2	200	72.7	W 16× 45	16⅛	—	7.4	11.4	144
9.0	15.9	194	70.6	W 12× 53	12	55.9	10.6	22.0	140
7.2	11.5	193	70.3	W 14× 48	13¾	—	8.5	16.0	139

ALLOWABLE MOMENT IN BEAMS
(C_b = 1, F_y = 36 ksi)

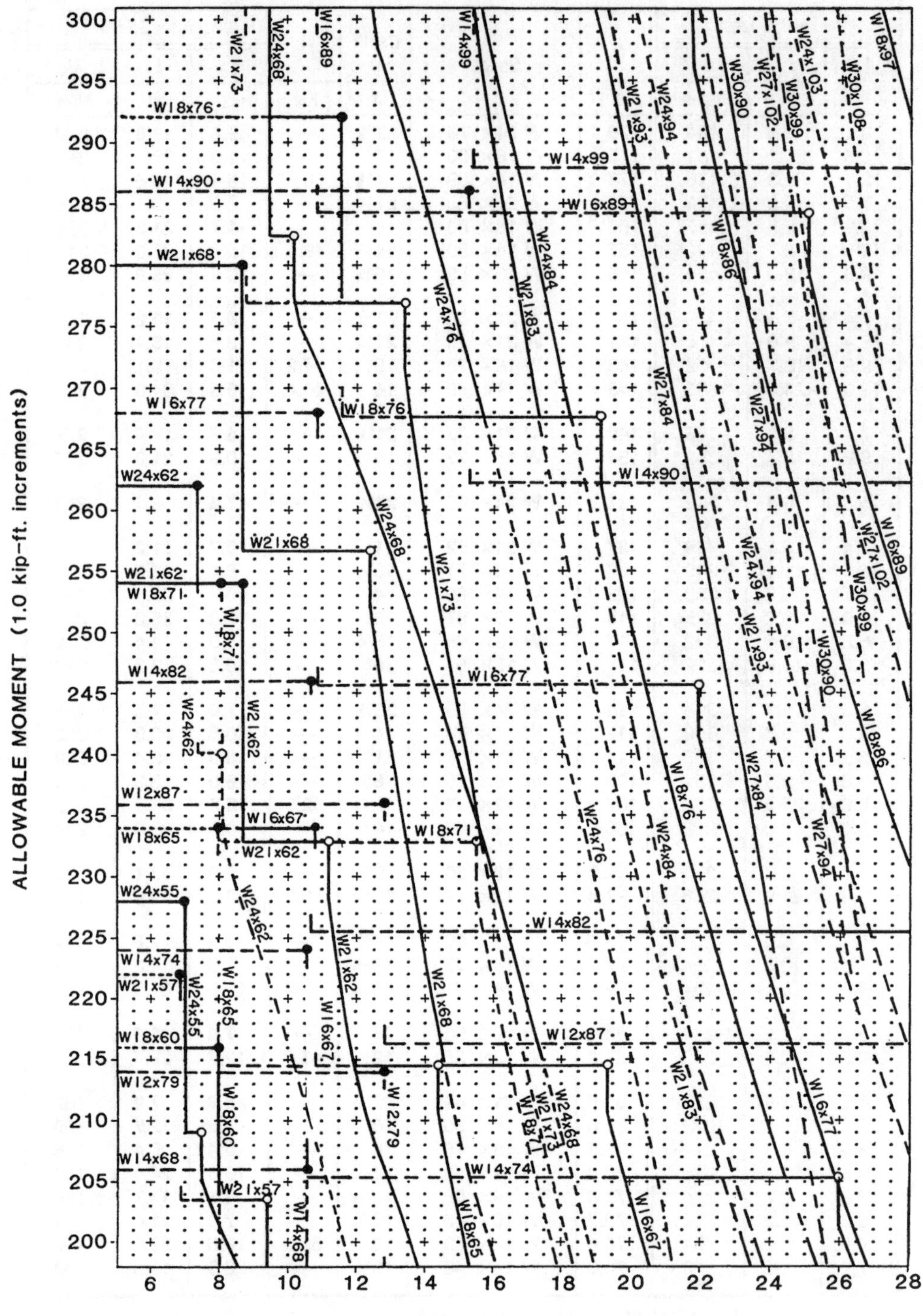

Reprinted from the *AISC Manual,* copyright © American Institute of Steel Construction, Inc. Reprinted with permission. All rights reserved.

TABLE 23-I-F—HOLDING POWER OF BOLTS[1,2,3] FOR DOUGLAS FIR-LARCH, CALIFORNIA REDWOOD (CLOSE GRAIN) AND SOUTHERN PINE (See U.B.C. STANDARD 23-17 WHERE MEMBERS ARE NOT OF EQUAL SIZE AND FOR VALUES IN OTHER SPECIES.)

p = safe loads parallel to grain, in pounds.
q = safe loads perpendicular to grain, in pounds.

× 4.45 for N

LENGTH OF BOLT IN MAIN WOOD MEMBER[4] (inches)		DIAMETER OF BOLT (inches)					
		3/8	1/2	5/8	3/4	7/8	1
		× 25.4 for mm					
1½	Single p	325	470	590	710	830	945
	Shear q	185	215	245	270	300	325
	Double p	650	940	1,180	1,420	1,660	1,890
	Shear q	370	430	490	540	600	650
2½	Single p		630	910	1,155	1,370	1,575
	Shear q		360	405	450	495	540
	Double p	710	1,260	1,820	2,310	2,740	3,150
	Shear q	620	720	810	900	990	1,080
3½	Single p			990	1,400	1,790	2,135
	Shear q			565	630	695	760
	Double p	710	1,270	1,980	2,800	3,580	4,270
	Shear q	640	980	1,130	1,260	1,390	1,520
5½	Single p					1,950	2,535
	Shear q					1,090	1,190
	Double p		1,270	1,990	2,860	3,900	5,070
	Shear q		930	1,410	1,880	2,180	2,380
7½	Single p						
	Shear q						
	Double p			1,990	2,860	3,890	5,080
	Shear q			1,260	1,820	2,430	3,030
9½	Single p						
	Shear q						
	Double p				2,860	3,900	5,080
	Shear q				1,640	2,270	2,960
11½	Single p						
	Shear q						
	Double p					3,900	5,080
	Shear q					2,050	2,770
13½	Single p						
	Shear q						
	Double p						5,100
	Shear q						2,530

[1]Tabulated values are on a normal load-duration basis and apply to joints made of seasoned lumber used in dry locations. See Division III for other service conditions.
[2]Double shear values are for joints consisting of three wood members in which the side members are one half the thickness of the main member. Single shear values are for joints consisting of two wood members having a minimum thickness not less than that specified.
[3]See Division III for wood-to-metal bolted joints.
[4]The length specified is the length of the bolt in the main member of double shear joints or the length of the bolt in the thinner member of single shear joints.

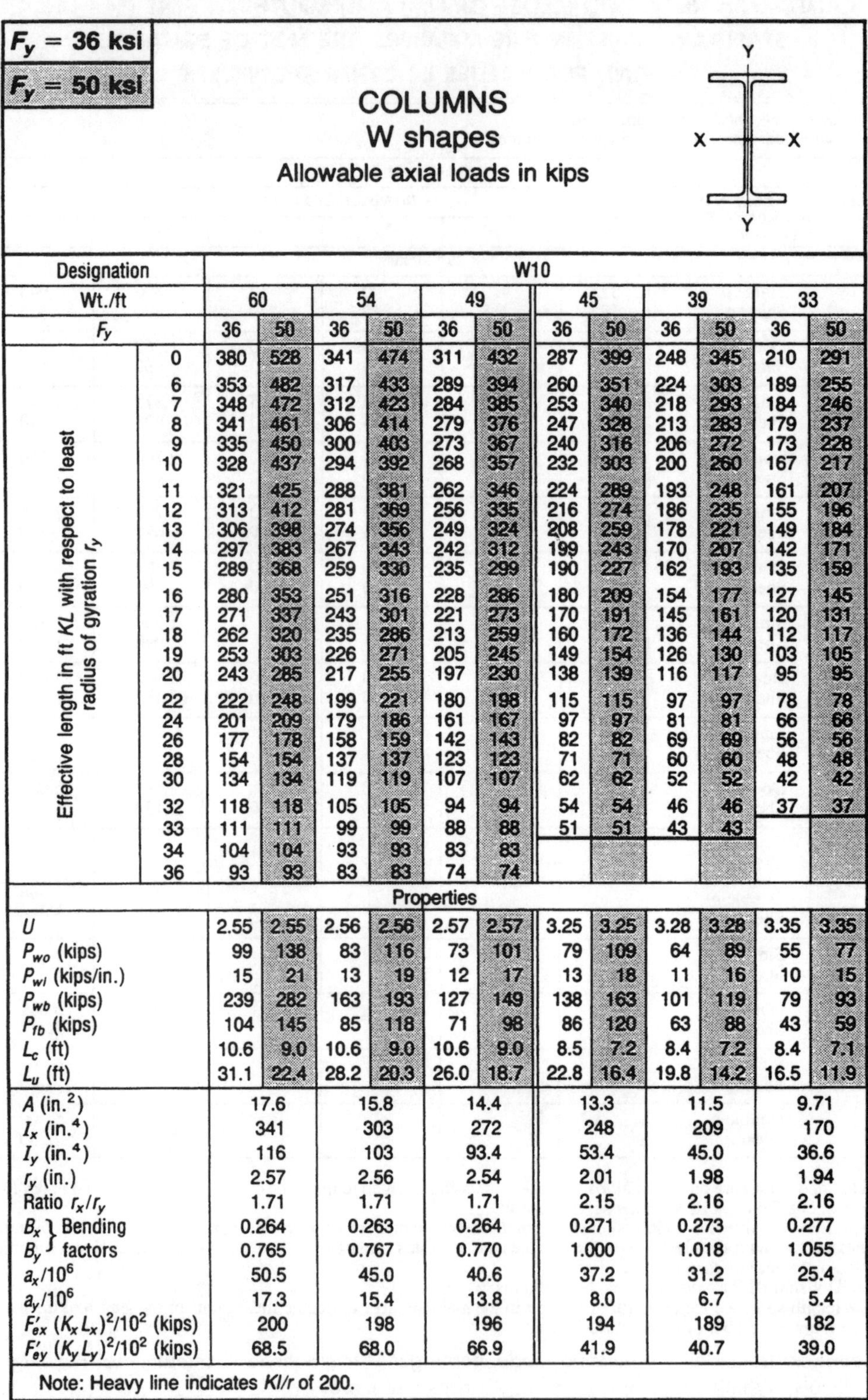

F_y = 36 ksi
F_y = 50 ksi

COLUMNS
W shapes
Allowable axial loads in kips

Designation		W10											
Wt./ft		60		54		49		45		39		33	
F_y		36	50	36	50	36	50	36	50	36	50	36	50
	0	380	528	341	474	311	432	287	399	248	345	210	291
	6	353	482	317	433	289	394	260	351	224	303	189	255
	7	348	472	312	423	284	385	253	340	218	293	184	246
	8	341	461	306	414	279	376	247	328	213	283	179	237
	9	335	450	300	403	273	367	240	316	206	272	173	228
	10	328	437	294	392	268	357	232	303	200	260	167	217
	11	321	425	288	381	262	346	224	289	193	248	161	207
	12	313	412	281	369	256	335	216	274	186	235	155	196
	13	306	398	274	356	249	324	208	259	178	221	149	184
	14	297	383	267	343	242	312	199	243	170	207	142	171
	15	289	368	259	330	235	299	190	227	162	193	135	159
	16	280	353	251	316	228	286	180	209	154	177	127	145
	17	271	337	243	301	221	273	170	191	145	161	120	131
	18	262	320	235	286	213	259	160	172	136	144	112	117
	19	253	303	226	271	205	245	149	154	126	130	103	105
	20	243	285	217	255	197	230	138	139	116	117	95	95
	22	222	248	199	221	180	198	115	115	97	97	78	78
	24	201	209	179	186	161	167	97	97	81	81	66	66
	26	177	178	158	159	142	143	82	82	69	69	56	56
	28	154	154	137	137	123	123	71	71	60	60	48	48
	30	134	134	119	119	107	107	62	62	52	52	42	42
	32	118	118	105	105	94	94	54	54	46	46	37	37
	33	111	111	99	99	88	88	51	51	43	43		
	34	104	104	93	93	83	83						
	36	93	93	83	83	74	74						

The left vertical label reads: *Effective length in ft KL with respect to least radius of gyration r_y*

Properties

		36	50	36	50	36	50	36	50	36	50	36	50
U		2.55	2.55	2.56	2.56	2.57	2.57	3.25	3.25	3.28	3.28	3.35	3.35
P_{wo} (kips)		99	138	83	116	73	101	79	109	64	89	55	77
P_{wi} (kips/in.)		15	21	13	19	12	17	13	18	11	16	10	15
P_{wb} (kips)		239	282	163	193	127	149	138	163	101	119	79	93
P_{fb} (kips)		104	145	85	118	71	98	86	120	63	88	43	59
L_c (ft)		10.6	9.0	10.6	9.0	10.6	9.0	8.5	7.2	8.4	7.2	8.4	7.1
L_u (ft)		31.1	22.4	28.2	20.3	26.0	18.7	22.8	16.4	19.8	14.2	16.5	11.9

	60	54	49	45	39	33
A (in.^{2})	17.6	15.8	14.4	13.3	11.5	9.71
I_x (in.^{4})	341	303	272	248	209	170
I_y (in.^{4})	116	103	93.4	53.4	45.0	36.6
r_y (in.)	2.57	2.56	2.54	2.01	1.98	1.94
Ratio r_x/r_y	1.71	1.71	1.71	2.15	2.16	2.16
B_x } Bending	0.264	0.263	0.264	0.271	0.273	0.277
B_y } factors	0.765	0.767	0.770	1.000	1.018	1.055
$a_x/10^6$	50.5	45.0	40.6	37.2	31.2	25.4
$a_y/10^6$	17.3	15.4	13.8	8.0	6.7	5.4
$F'_{ex}(K_xL_x)^2/10^2$ (kips)	200	198	196	194	189	182
$F'_{ey}(K_yL_y)^2/10^2$ (kips)	68.5	68.0	66.9	41.9	40.7	39.0

Note: Heavy line indicates Kl/r of 200.

BEAM DIAGRAMS AND FORMULAS
For various static loading conditions

For meaning of symbols, see page **2 - 293**

---

1. SIMPLE BEAM—UNIFORMLY DISTRIBUTED LOAD

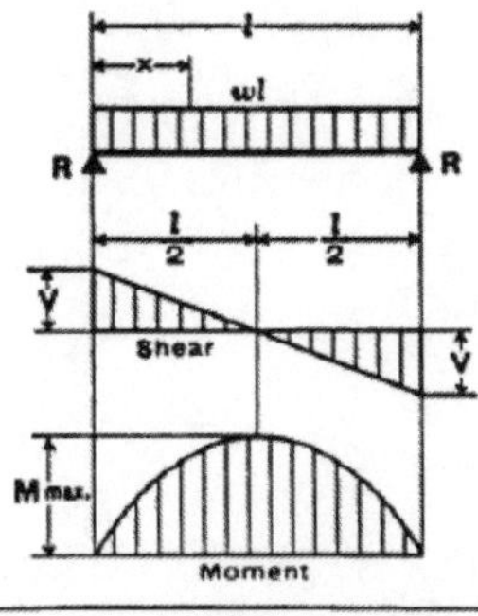

Total Equiv. Uniform Load . . . . $= wl$

$R = V$ . . . . $= \dfrac{wl}{2}$

V_x . . . . $= w\left(\dfrac{l}{2} - x\right)$

M max. $\left(\text{at center}\right)$ . . . . $= \dfrac{wl^2}{8}$

M_x . . . . $= \dfrac{wx}{2}(l - x)$

Δmax. $\left(\text{at center}\right)$ . . . . . $= \dfrac{5\,wl^4}{384\,EI}$

Δ_x . . . . $= \dfrac{wx}{24EI}(l^3 - 2lx^2 + x^3)$

---

7. SIMPLE BEAM—CONCENTRATED LOAD AT CENTER

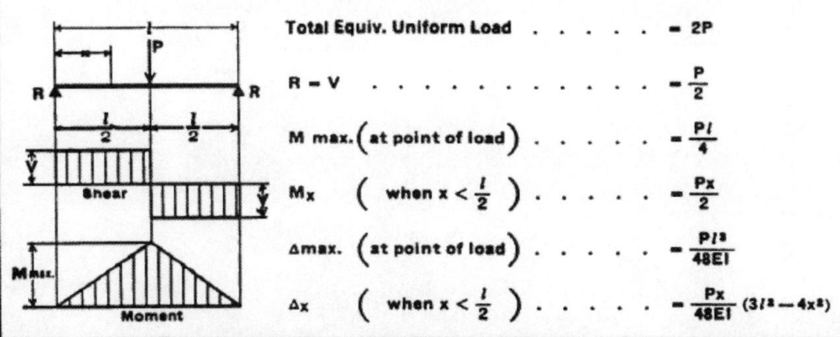

Total Equiv. Uniform Load . . . . . . $= 2P$

$R = V$ . . . . $= \dfrac{P}{2}$

M max. $\left(\text{at point of load}\right)$ . . . . . $= \dfrac{Pl}{4}$

M_x $\left(\text{when } x < \dfrac{l}{2}\right)$ . . . . . $= \dfrac{Px}{2}$

Δmax. $\left(\text{at point of load}\right)$ . . . . . $= \dfrac{Pl^3}{48EI}$

Δ_x $\left(\text{when } x < \dfrac{l}{2}\right)$ . . . . . $= \dfrac{Px}{48EI}(3l^2 - 4x^2)$

---

22. CANTILEVER BEAM—CONCENTRATED LOAD AT FREE END

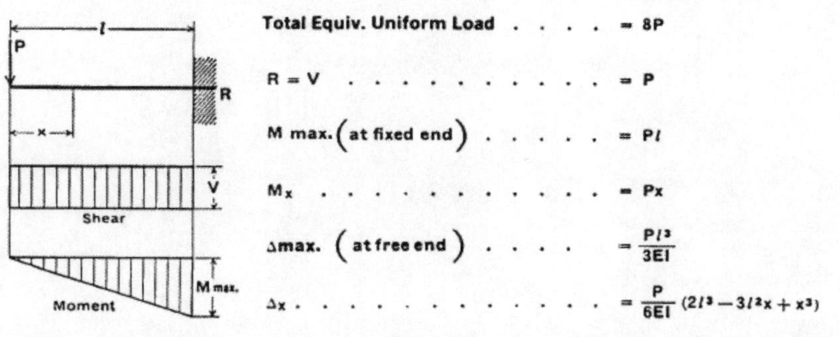

Total Equiv. Uniform Load . . . . . $= 8P$

$R = V$ . . . . $= P$

M max. $\left(\text{at fixed end}\right)$ . . . . . $= Pl$

M_x . . . . $= Px$

Δmax. $\left(\text{at free end}\right)$ . . . . $= \dfrac{Pl^3}{3EI}$

Δ_x . . . . $= \dfrac{P}{6EI}(2l^3 - 3l^2x + x^3)$

1.   **C.** The maximum moment of the girders in either plan equals $wL^2/8$, where w is the uniform load, which is the same for both plans, and L is the girder span. Thus, the maximum moment in the plan A girders = $wL^2/8 = 72w$, and the maximum moment in the plan B girders = $wL^2/8 = 288w$, or four times as great ($288/72 = 4$). As the required section modulus is equal to the maximum moment divided by the allowable flexural stress ($S = M/F_b$), the required section modulus of the plan B girders is four times as great as that of the plan A girders.

2.   **D.** Three types of stress that are important in building design are tension, compression, and shear (A, B, and C). Strain (D) is not a type of stress; it is the deformation, or change in size, of a body caused by external loads.

3.   **D.** Structural steel columns tend to fail by buckling. The slenderness ratio KVr is a measure of the buckling tendency of a steel column; the larger the value of KVr, the greater the tendency of the column to buckle, resulting in a lower column capacity. In this ratio, K is a constant determined by the end conditions of the column; that is, whether the column is free to translate (move laterally). The unbraced length of the column (A) is 1, and the moment of inertia and area of the column (B) determine its radius of gyration r. The grade of steel (B) does not affect the value of K.

4.   **D.** All four methods will strengthen the beam; what we are looking for is the most effective method. How do we approach this problem? As no information is given about loads, span, or beam size, the only fact we know for sure is that we want to increase the beam's flexural strength. To accomplish this, we must increase its section

modulus $S(= I/c)$. To increase the section modulus efficiently, we must provide as much material as possible the maximum distance away from the neutral axis. Of the four choices offered, D most effectively satisfies this concept.

5.   **B.** It is important for exam candidates to be familiar with shear and moment diagrams. Looking at the four choices in this question, A is the moment diagram for a simple beam with a concentrated load at the center; B is the moment diagram for a beam fixed at both ends with a concentrated load at the center, which is the case in this question; C is the shear diagram for a beam (simple or fixed at both ends) with a concentrated load at the center; and D is the moment diagram for a beam fixed at both ends with a uniformly distributed load.

6.   **D.** The shear stress in a column pad is essentially a function of the column load, the column size, and the thickness of the pad. It is not at all related to the reinforcing steel (B and C), and only slightly affected by the pad size (A). Increasing the pad thickness (correct answer D) provides more area of concrete to resist the shear load, and hence decreases the shear stress.

7.   **A.** The flexural stress in a homogeneous rectangular beam varies from zero at the neutral axis to a maximum value at the top and bottom fibers, as shown in correct answer A. C indicates the variation of shear stress, and D shows the stress in a reinforced concrete beam at failure.

8.   **D.** All three conditions described will cause bending in truss members (D). The members of a truss without diagonals (I), known as a Vierendeel truss, are subject to bending moment in addition to axial forces. The use of closely spaced joists (II) loads

the trusses between panel points, which results in bending moment in the truss chord members that are loaded. In truss analysis, the joints are generally assumed to be hinged. In actual practice, however, the joints are not hinged; the truss members are restrained against rotation by the gusset plates to which they are connected. This restraint introduces small bending stresses, known as secondary stresses.

9. **B.** It is not necessary to memorize any deflection formulas in order to answer this question. But, at least, you should know that deflection formulas generally are in this form: Deflection $= KWL^3/EI$, where K depends on the loading, W is the total load, L is the length, E is the modulus of elasticity, and I is the moment of inertia. Therefore, to decrease the deflection by 20 percent ($0.90/0.75 = 1.20$), one would increase the moment of inertia (I) by 20 percent (correct answer B).

10. **B.** The ultimate tensile capacity of a reinforcing bar is the product of its area and yield strength ($T_u = A_s f_Y$). Grade 40 reinforcing steel has a yield strength of 40 ksi, while grade 60 has a yield strength of 60 ksi. For 2-#6, using the table given in the problem,

$A_s = 0.44$ in.$^2 \times 2$ bars $= 0.88$ in.^{2}

$T_u = 0.88$ in.$^2 \times 40$ kips/in.$^2 = 35.2$ kips

For 1-#8,

$A_s = 0.79$ in.^{2}

$T_u = 0.79$ in.$^2 \times 60$ kips/in.$^2 = 47.4$ kips

Therefore B is correct. D is incorrect; although one must know the concrete strength to determine the capacity of a beam, the tensile capacity of the bars themselves is independent of the concrete strength.

11. **A.** A flat slab floor is a two-way reinforced concrete system in which the supporting beams are eliminated and the slab is supported directly on the columns. A flat plate floor is a special type of flat slab that does not have a thickened slab at the columns or an enlarged section at the tops of the columns. The floor slab of uniform thickness is thus supported directly by columns of uniform cross-section. Statement IV is therefore incorrect. Flat plate floors are often economical where the spans are moderate and the loads relatively light (statement V is incorrect). Construction depth for each floor is minimum (correct statement I). Some problems inherent in the flat plate system include high shear stresses in the slab near the columns (correct statement II), and high deflection of the relatively thin slab (incorrect statement III). To sum up, only statements I and II are correct (correct choice A).

12. **D.** Continuous beams (those that rest on more than two supports) offer certain advantages over simple beams. The maximum positive bending moment in a continuous beam is less (A), and the maximum deflection is also less (incorrect statement D). However, a continuous beam is subject to negative bending moment over its supports, while a simple beam never has negative moment (B). C is also correct: for equal spans and loads, the maximum positive bending moment is greater in the end spans of a continuous beam than in the center span.

13. **B.** This question tests your understanding of beam behavior. In a simple beam subject to downward vertical loads, the upper beam fibers (1) shorten and are in compression,

while the lower beam fibers (3) lengthen and are in tension. The fibers at mid-depth of the beam (2) do not change in length. The correct answer is therefore B.

14. **B.** This problem may be solved either analytically or graphically. In the analytical solution, we first determine the truss reactions.

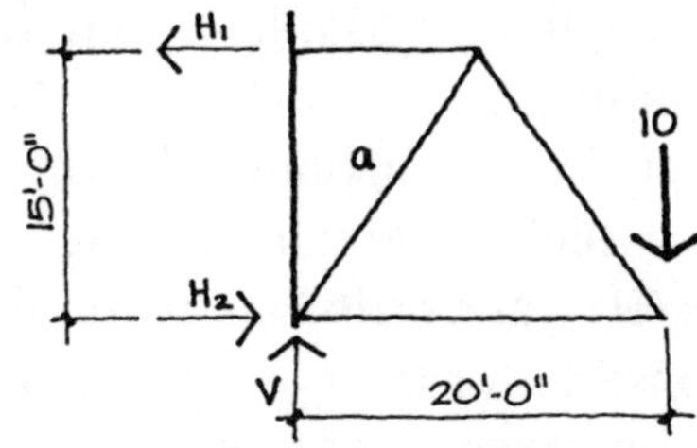

$$\Sigma M = 0$$

$$10(20) - H_1(15) = 0$$

$$H_1 = \frac{10\ (20)}{15} = 13.33 \text{ kips}$$

$$\Sigma H = 0 \quad H_2 - H_1 = 0$$

$$H_1 = H_2$$

$$H_2 = 13.33 \text{ kips}$$

$$\Sigma V = 0$$

$$V - 10 = 0$$

$$V = 10.0 \text{ kips}$$

Next we isolate the truss joint at the lower reaction.

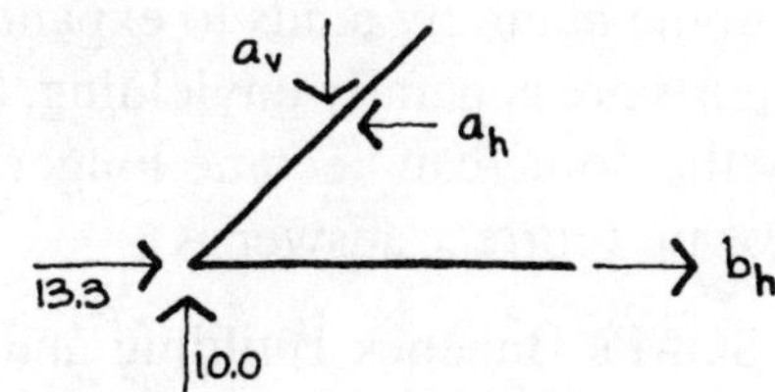

$$\Sigma V = 0 \quad 10 - a_v = 0 \quad a_v = 10.0 \text{ kips}$$

Because the horizontal projection of member a is 10 feet and its vertical projection is 15 feet,

$$a_h = \frac{10}{15} \times a_v = \frac{2}{3} \times 10.0 = 6.67 \text{ kips}$$

The total stress in member a is

$$\sqrt{(a_v)^2 + (a_h)^2} = \sqrt{(10.0)^2 + (6.67)^2}$$

$$= 12.02 \text{ kips}$$

Because the stress acts *towards* the joint, the stress is *compressive*. The correct answer is therefore B.

15. **B.** Standardized weld symbols are used to convey information about the required welding. The weld symbol approximates the shape of the weld; in this case ⊽ indicates a fillet weld. If a bevel weld (B) were specified, the symbol ⊾ would have been shown. The number to the left of the weld symbol indicates the size of the weld, in this case 1/4 inch. The number to the right of the weld symbol indicates the length of the weld if one number is shown, or the length and spacing if two numbers are shown, in this case 2-inch long welds at 12 inches on center. The darkened flag indicates that the weld is to be made in the field, as opposed to the shop. Information on weld testing and electrodes is not usually shown on the weld symbol.

16. **D.** Formwork represents a large portion of the cost of any reinforced concrete building, and therefore the use of repetitive and reusable forms is one of the most important steps in achieving an economical concrete building. In addition, having a simple and repetitive reinforcing steel layout is also economical (D is correct). The amount of reinforcing steel is important, but not so much so that the concrete sizes are increased disproportionately (A), As this adds to the weight of the building, increases footing sizes, and may have adverse seismic implications. In general, for low-rise buildings, vertical and lateral loads are economically resisted by bearing/ shear walls, whereas

in high-rise buildings, rigid frames are often more economical. Hence, B is not correct. While C correctly describes the advantages of waffle slab construction, this is not always the most economical system, and thus C is not the correct answer.

17. **A.** To solve for the left reaction, we take moments about the right reaction.

$$\Sigma M = 0$$
$$-5,000 \ (24 + 8) \ (16) - 12,000 \ (24 - 8) -$$
$$5,000 \ (32) + 24R_L = 0$$
$$R_L = \frac{5,000(32)(16) + 12,000(16) + 5,000(32)}{24}$$

25,333# (answer A)

18. **D.** All of the terms in this question are associated with Roman architecture, except *stoa* (correct answer D). *Thermae* were Roman public bathing establishments, a *basilica* was a rectangular building used as a hall of justice and public meeting place, and an *amphitheater* was an oval arena surrounded by tiers of seats and used in Rome for gladiatorial contests and other events. A *stoa* was a colonnade used by the ancient Greeks to provide access to law courts, gymnasiums, and so forth.

19. **A.** Because the load decreases uniformly, the slope of the shear diagram also decreases, and the shear at the left support is greater than at the right support. These conditions are correctly shown in diagram A.

20. **C.** This question tests your knowledge concerning one of the great structures of the Renaissance. All of the statements are correct, except C. The dome contains a number of circular compression rings, which made it stable during construction without the use of any temporary shoring, an incredible achievement for its time.

21. **D.** The Pantheon, built in the year 123, remains to this day one of the greatest achievements of Roman architecture. Its immense concrete dome is so thick at its bottom that tensile hoop stresses are resisted by the concrete, without any need for iron reinforcing. The opening at the crown of the dome was designed and built to act as a compression hoop.

22. **D.** All of the men in this question, with the exception of Maillart, were important figures in the development of the sky-scraper. Otis (I) invented the first safe elevator, which made tall buildings practical. Khan (II) was a brilliant structural engineer responsible for the design of two of the most important skyscrapers of modern times: the John Hancock Building and the Sears Tower, both in Chicago. Jenney (III) designed the first skyscraper, the Home Insurance Co. Building of 1883. Maillart (IV), a Swiss engineer, was noted for his arched concrete bridges.

23. **B.** The roof of the Dulles Airport terminal is supported by a series of steel cables suspended between huge concrete piers. The concept of leaning these piers outward may have been inspired by aesthetic considerations, or to express the idea of flight (A), but the main purpose was to counteract the inward pull of the cables (B).

24. **A.** When the outside temperature goes up, the dome naturally tends to expand. As its supports are generally unyielding, the only way the dome can become longer is to move up (correct answer A).

25. **C.** SOM's Hancock Building and Sears Tower are tubular buildings, in which the perimeter walls form an immense hollow tube that cantilevers out of the ground under the action of wind loads. The Sears Tower is actually a bundle of nine such tubes, which terminate at varying heights.

Both buildings, among the tallest in the world, are framed in structural steel. As I and II are both correct, C is the correct answer.

26. **B.** The correct definition of live load is given in answer B. There is no relationship between dead and live load (A), live load may not be neglected unless specifically permitted by the building code (C), and although long-span structures are often more vulnerable to failure than conventional structures, there is no requirement that they be designed for a greater live load (D).

27. **C.** Certain types of structures obtain their strength by their shape. Such form-resistant structures include membranes, which resist loads by tension, and thin shells, which develop compression, shear, and tension.

28. **D.** An arch is a long-span structure whose internal stresses are essentially compressive. A three-hinged arch has hinges at each support and at the top, or crown. The horizontal thrust at each support is directly proportional to the load (I) and the span (II), and inversely proportional to the rise (III is incorrect). A three-hinged arch rotates when the temperature changes, rather than undergoing any change in the horizontal thrust (IV is incorrect). The correct answer is therefore D.

29. **C.** The behavior of long-span structures is not exactly the same as that of conventional structures. Secondary effects may become critical, and consequently safety factors may need to be increased (A). Because the levels of quality control of site and factory-assembly are not the same, the architect should consider using different factors of safety (B). The effects of temperature are more pronounced in long-span structures because of the greater length of members (D). However, C is incorrect, and therefore the answer to this question. Building codes establish minimum criteria for design, but the architect may choose to use higher factors of safety for a particular structure.

30. **A.** A membrane is a thin, flexible sheet that can only resist tension (correct answer A). Examples include circus tents, umbrellas, and trampolines. Some permanent structures, too, have spanned great distances with the use of membranes, most notably the Haj Terminal in Saudi Arabia.

31. **A.** An air-supported roof is a membrane pretensioned by internal air pressure, enabling it to resist downward loads without wrinkling or buckling. In this case, the uplift from the wind increases the tension in the membrane. The pull of the tensed membrane on the concrete foundation ring is tangent to the surface of the membrane; its components therefore act upward and inward. The upward component is resisted by the weight of the concrete ring, while the inward component tends to shorten the concrete ring, thus stressing it in compression (correct answer A).

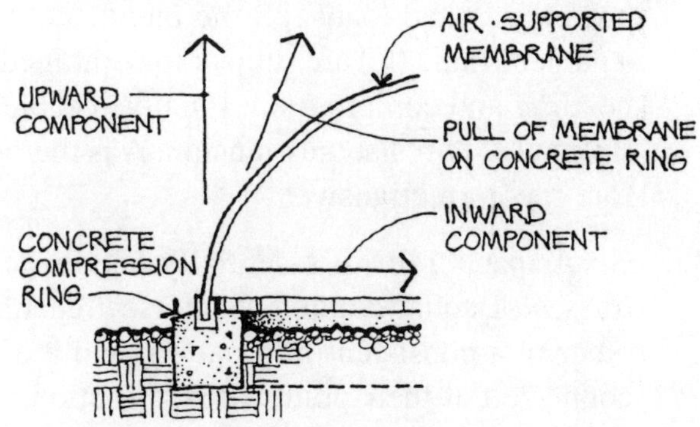

32. **B.** Candidates should be conversant with some of the terminology of long-span structural systems. The groined vault (A)

is a structure devised by the ancient Romans in which two barrel vaults intersect at right angles. The Schwedler dome (C) is a type of steel-framed dome, comprising meridional ribs and circular hoops. The hyperbolic paraboloid (D) is a type of thin shell structure whose surface shape is determined by moving a parabola with downward curvature along a perpendicular parabola that is curved upward. The correct answer is lamella (B).

33. **D.** A shell, or thin shell, structure has a curved surface and supports load by tension, compression, and shear in its own plane (II is correct, IV is incorrect). However, it is too thin to resist bending stresses (III is correct), which makes it unable to support any substantial concentrated loads (I is incorrect). The correct statements are II and III (answer D).

34. **A.** An arch (A) is primarily a compression structure, although it may also be subject to some bending stress. The horizontal members of a rectangular rigid frame (B) are stressed principally in bending, plus some axial compression, while the vertical members are stressed primarily in compression, with some bending. When supporting vertical loads, all the members of a gabled frame (C) are subject to combined bending and compression. Finally, a cable (D) is able to resist only tension. A is therefore the correct answer.

35. **C.** A space frame is essentially a series of trusses of equal depth that intersect each other in a consistent grid pattern and are connected at their points of intersection. Loads are supported by the trusses in both directions, and the entire system works as a unit. Sometimes, the top and bottom chords in both directions are offset by half a module, creating inclined, rather than vertical, trusses.

36. **C.** A cable only resists tension, and tends to pull away from its supports. The radial cables pulling away from the inner ring tend to open it up or stretch it, resulting in tension in the ring. Similarly, the cables tend to close or compress the outer ring. C is therefore the correct answer.

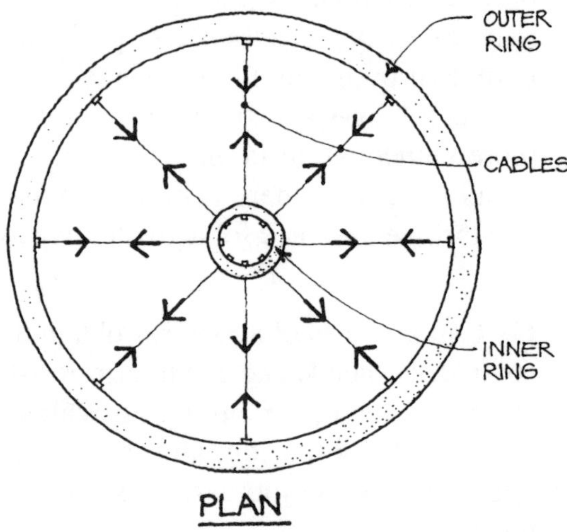

37. **C.** The truss without diagonals in this question is known as a Vierendeel truss, although it is actually a type of rigid frame. If its joints are rigid and capable of resisting moment, it is a stable structure (B and D are incorrect). As there are no diagonals, the chord members must resist shear, which produces bending moments in the chords. These moments are greatest where the maximum vertical shear occurs, near the supports (correct statement C). Vierendeel trusses tend to be uneconomical in the use of material, and also have high deflection (A is incorrect).

38. **A.** Long-span structures are used for a variety of reasons and a variety of buildings. For maximum flexibility, columns in exhibition halls (I) should be spaced as far apart as possible. Basketball pavilions (II), auditoriums (IV), and other places of public entertainment and assembly require large column-free areas for optimum visibility and flexibility. The other building types (library, university science building, and municipal administration building) generally do not require large column-free flexible spaces, and therefore are not likely candidates for long-span construction. A is the correct answer.

39. **A.** A hyperbolic paraboloid is a saddle-shaped surface formed by moving a parabola with downward curvature along a perpendicular parabola that is curved upward. Although its surface is doubly curved, it can be formed using a series of straight timber planks.

40. **B.** Where two vaults perpendicular to each other intersect in Gothic cathedrals, the resulting form is called a groined vault. The diagonal lines of intersection, called groins, transmit the load to piers at the four corners.

41. **C.** All suspension structures have a cable as their essential structural element. Cables can resist only tension, and are always curved upward.

42. **D.** In a rigid frame, the joint between its beam and column members is rigid; that is, the angle between the members cannot change, and bending moment can be transferred. This enables the rigid frame to resist both horizontal and vertical loads.

43. **A.** B is the deflected shape of a simple beam supported by columns, without rigid joints. C is the deflected shape of a rigid frame with fixed supports, while D is the shape the hinged frame would take if horizontal reactions could not be resisted at the bases. A is the correct answer.

44. **C.** A truss, one of the oldest methods employed for long-span construction, is a structure designed to support vertical or horizontal loads and composed of straight members that form a number of triangles. Trusses may have a variety of configurations, but must always consist of triangles.

45. **C.** A three-hinged arch is statically determinate (I); that is, its reactions can be determined using only the equations of equilibrium. II is also correct. However, III is incorrect; the hinge at the center does not allow any moment to be developed. C is therefore the correct answer.

46. **C.** If no strain remains when the stress is removed, it is called elastic action. However, if some strain remains, the action is termed inelastic. Steel is elastic up to the elastic limit and inelastic at stresses above the elastic limit. Concrete and wood are nearly elastic at low stresses, but inelastic at higher stresses.

47. **B.** If a question like this were to appear on the exam, it is likely that the dimensions of the column would be found in tables reproduced from the AISC Manual, rather than in the question itself. In any event, one should know which information is relevant and which is not. In this case, the compressive stress is simply the load P divided by the area A. All the other information given is irrelevant. Therefore, the compressive stress = P/A = 200 kips/14.1 in.^{2} = 14.2 ksi.

48. **B.** Exam questions often test candidates' understanding of structural concepts, as in this question. For a given load and span, a simple beam (A and D) will generally have the greatest deflection and positive moment. In arrangement B, beam 1-2 extends past support 2 and supports the end of the simple beam in span 2-3. This has the effect of reducing both the positive moment and the deflection in span 1-2. In arrangement C, beam 2-3 extends to about the midpoint of span 1-2 and will be subject to high negative moment at support 2, as well as high deflection. The correct answer is therefore B.

49. B. When purlins or other framing members are located at truss panel points, the chord and web members of the truss are stressed in axial tension or compression only and have no bending stress. However, if load is applied to a truss chord between panel points, the chord acts as a beam spanning between the panel points and is thus subject to bending stress in addition to the axial tension or compression caused by the truss action.

50. B. For elastic materials, such as steel, unit stress is proportional to unit strain at stresses below the elastic limit. If the stress is doubled, the strain is doubled, and so on. The stress-strain diagram in this region is therefore a straight line. The ratio of unit stress to unit strain is called the modulus of elasticity and is designated E. Materials with a high value of E are said to be stiff, which means highly resistant to deformation.

51. C. This truss resists both a vertical load and a horizontal load. Notice that support B is on a roller, which means that there can be no horizontal reaction at that support. Therefore, there must be a horizontal reaction of 10,000# acting to the right at support A, from $\Sigma H = 0$. To determine the vertical reaction at A, we take moments about B.

$$\Sigma M_B = 0$$

$$10,000(10 \text{ ft.}) - 10,000(10 \text{ ft.}) + V_A(40 \text{ ft.}) = 0$$

$$V_A = 0$$

From $\Sigma V = 0$, the vertical reaction at B equals 10,000# upward. The reactions are shown correctly in choice C.

52. D. Rigid frames are widely used in building construction, and candidates should therefore understand their basic behavior. The rigid beam-column joints restrain the ends of the beam, resulting in negative moment in the beam at the columns and reduced positive moment between the columns. There is moment in each column, which varies from zero at the hinged base to a maximum value at the beam. D is therefore correct. Choice A shows the moment diagram if the beam were simply supported and not part of a rigid frame, B is the moment diagram if a concentrated horizontal load acted on the frame, and C correctly shows the beam's moment diagram but omits the moment diagram for the columns.

53. A. Ductility, as correctly defined in this question, is especially desirable for systems that resist earthquake loads. Strength (B) refers to the ability to resist load. Yielding (C) is deformation that a material undergoes at a certain level of stress, with no increase of load. Stiffness (D) means resistance to deformation.

54. B This problem tests candidates' understanding of continuous beams and may be approached non-mathematically as follows: beam AB deflects downward under load, and if there were no reaction at C, the beam would lift off of C as shown.

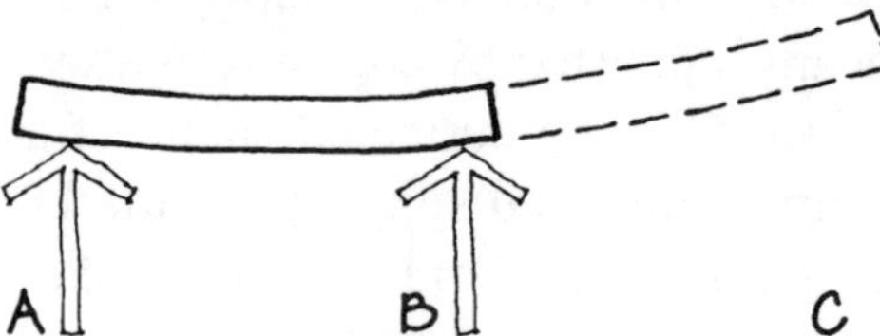

But because the beam is connected to C, the reaction at C pulls *downward* on the beam to prevent it from lifting off.

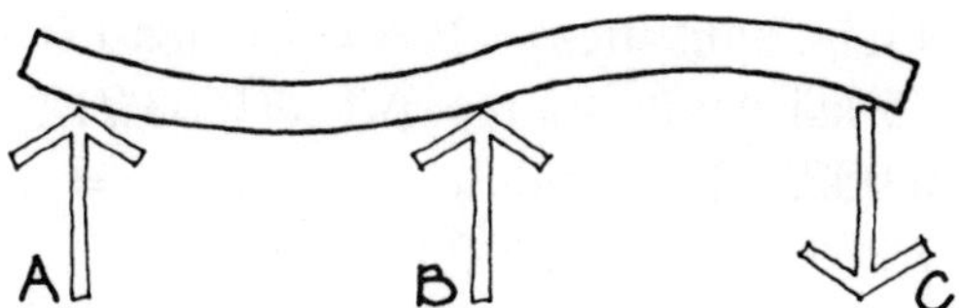

One can also solve this problem analytically. There is negative moment at support B (tension at the top, compression at the bottom), since B is an interior support of a continuous beam. We isolate span BC and take moments about B, assuming that the reaction at C is downward.

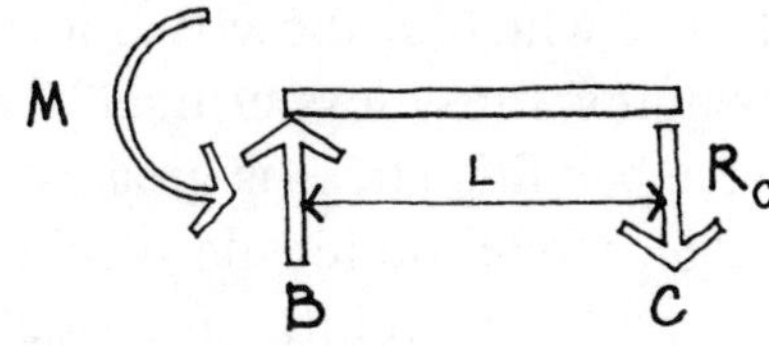

$\Sigma\, M_B = 0$

$-M + R_C(L) = 0$

$R_C = +M/L$

Because R_C comes out positive, our assumption that it acts downward is correct.

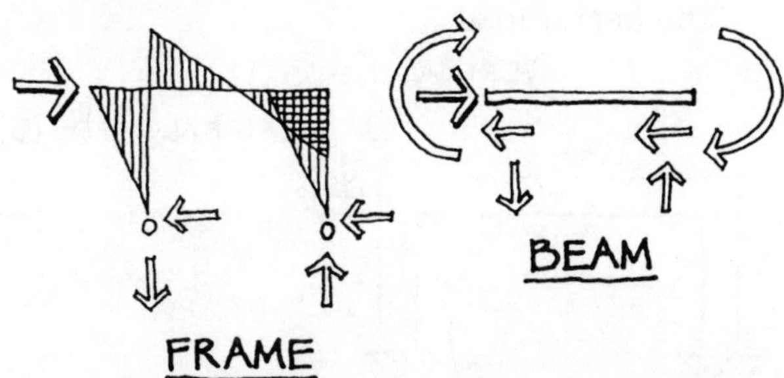

55. C. Questions of this type can be frustrating because the differences among the deflected shapes are subtle. As the column bases are hinged, they provide no restraint against rotation. A and C correctly show the column base condition. If the column bases were fixed, they would be restrained against rotation, as shown in B and D. The lateral load causes moment in the frame members as shown. The beam therefore distorts as shown in C and D. Only choice C correctly shows both the column and beam conditions.

56. A. Candidates are expected to be able to solve simple problems in statics, as in this question.

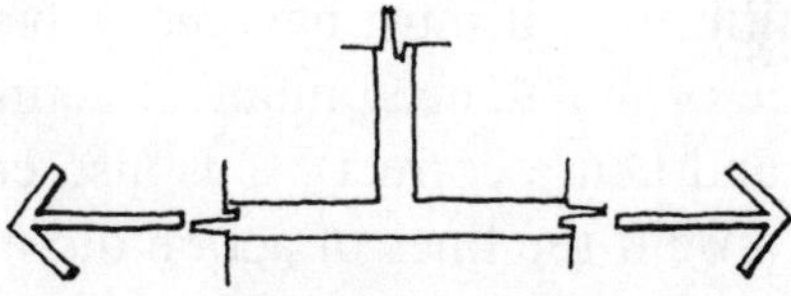

We isolate the bar. As member DB is a horizontal cable, it can only be stressed in tension. Therefore, it pulls away from the bar (to the left) in a horizontal direction. As $\Sigma\, H = 0$, H_A must be equal and opposite to H_B and act to the right. From the basic equation $\Sigma\, V = 0$, $-1,000 + V_A = 0$. $V_A = +1,000$, which means that V_A acts upward.

57. A. This problem can be solved very quickly if it is approached correctly. We simply isolate the joint between the vertical and bottom chord and apply the equation $\Sigma\, V = 0$.

As there are no vertical loads applied at the joint, the internal force in the vertical member is zero.

58. B. To answer this question, you simply need to know two things: (1) that a force that pulls on a member causes tension, and (2) that the grain of a wood member runs parallel to its length. Therefore, each 2×8

wood member is loaded in tension parallel to the grain.

59. C. The load applied to each vertical support is 50#/lin. ft. × 6'-0" = 300#. We solve for the force in the upper bolt by taking moments about the lower bolt.

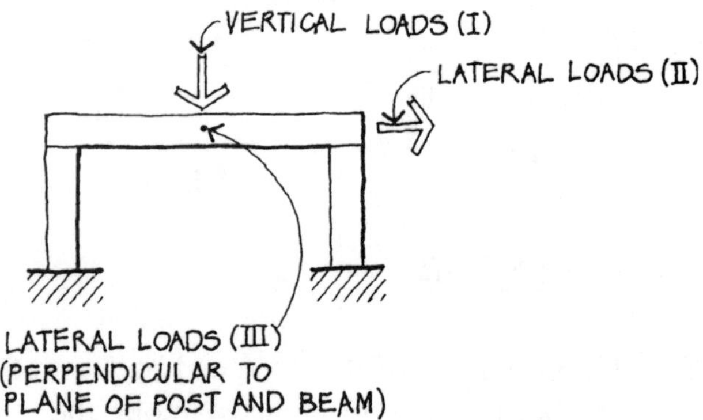

$$\Sigma M = 0$$

$$-300(3.75 + 1) + \text{Bolt force}(1) = 0$$

$$\text{Bolt force} = (300)(4.75)/1 = 1,425\#$$

As the force in the bolt acts *away* from the vertical support, the bolt is in tension.

60. B. One of the basic concepts in structures is that of equilibrium. For a body to be in equilibrium, it must have no unbalanced force or unbalanced moment acting on it (A and D are correct). C is also correct, because if the lines of action did not meet at a point, it would result in an unbalanced moment. B is the incorrect statement: if the forces acting on a body are balanced, this does not necessarily mean that the body is in equilibrium. The moments must also be balanced, or the body would rotate.

61. D. Rather than trying to memorize a formula, it is probably easier to remember that the modulus of elasticity of a material is simply the unit stress divided by the unit strain. Unit stress = P/A = 75 kips/1.0 in.^{2} = 75 ksi. Unit strain = Δ/L = 0.03"/12" = 0.0025. The modulus of lasticity = 75/0.0025 = 30,000 ksi.

62. C. An opening in a beam has the least effect on the beam's load-carrying capacity if it is located where both the shear and bending stresses are low. The shear stress is usually greatest at the supports (A and B) and least near midspan (C and D). Thus, considering shear stress, an opening at C or D would least affect the beam's load-carrying capacity. But at which of these two locations is the bending stress less critical? The compressive bending stress is greatest near the top edge, while the tensile bending stress is resisted by the reinforcing steel, which is close to the bottom edge (D). Therefore, the only location where both the shear and bending stresses are low is C.

63. A. A post and beam system can only resist vertical loads (I). It cannot resist lateral loads either in its own plane (II) or perpendicular to its own plane (III). Some type of bracing must be provided to resist the lateral loads.

64. C. Candidates should be able to determine the shears and moments in a simple beam subject to any type of loading, as in this question. We first calculate the left reaction by taking moments about the right end.

$\Sigma M = 0$

$-1.0(20)(20/2) - 20(6) - 10(8 + 6) + R_L(20) = 0$

$R_L = (200 + 120 + 140)/20 = 460/20$

$= 23.0$ kips

We next determine the point of zero shear.

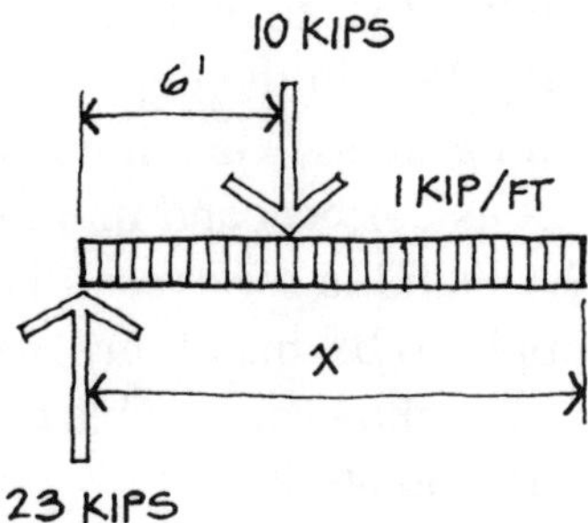

$+23.0 - 10.0 - 1.0(\times) = 0$

$\times = (23.0 - 10.0)/1.0 = 13.0$ ft.

Finally, because we know that the moment is maximum where the shear is zero, we calculate the moment at that point.

$M = +23.0(13.0) - 1.0(13.0)(13.0/2)$
$\quad -10.0(13.0 - 6.0) = 299.0 - 84.5 - 70$
$\quad = 144.5$ ft.-kips

65. D. This is a very simple problem, particularly as we're given the formula I = $bd^3/12$. In the formula, b = width = 4" and d = depth = 12". Thus, I = $4(12)^3/12 = 576$ in.^{4}.

66. C. Dead load is the vertical load due to the weight of all permanent structural and non-structural components of a building. Live load is the vertical load caused by the use and occupancy of a building, not including wind, earthquake, or dead loads. Building structures must be designed to resist various combinations of loads, the most basic of which is dead load plus live load.

67. C. We construct a force triangle, consisting of the 1,000# load, the tension in cable A, and the tension in cable B.

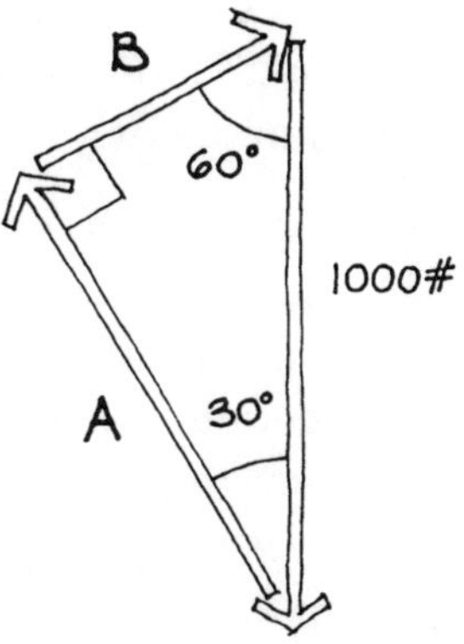

From basic trigonometry, sin 60° = A/1,000. A = 1,000 sin 60° = 1,000(0.866) = 866#. sin 30° = B/1,000. B = 1,000 sin 30° = 1,000(0.500) = 500#.

68. A. Hydrostatic pressure, which is the pressure exerted by a liquid, is equal to the unit weight of the liquid (I) multiplied by the depth of the liquid (II). The width of the tank that contains the liquid (III) is irrelevant.

69. D. An arch is primarily a compression member. In fact, under a given loading, it may be stressed purely in compression. Any other loading will result in some bending, in addition to compression. Therefore, in general, an arch supports load by a combination of compression and bending.

70. D. In this problem, we must determine the magnitude of the horizontal thrust. Taking moments about the crown,

$VL/2 - (wL/2)(L/4) - Hh = 0$.
$Hh = (wL/2)(L/2) - wL^2/8 = wL^2/8$.
$H = wL^2/8h$.
Therefore, if w and h remain the same, but L becomes 2L, the new value of H = w

$(2L)^2/8h = 4wL^2/8h$, which is 4 times the original value of H.

71.   A. Shear stress in a rectangular homogeneous beam varies parabolically, from zero at the top and bottom fibers to a maximum value at the neutral axis.

72.   B. A hinge or pin is free to rotate and therefore has no resistance to moment. Actual hinges or pins are rarely used in modern construction. However, a joint that has little moment resistance or restraint against rotation is generally considered to be a hinge or pin.

73.   A. In problems of this kind, we cut an imaginary section through the truss that cuts through the member whose internal force is to be determined. We then draw a free body diagram and apply the equations of equilibrium.

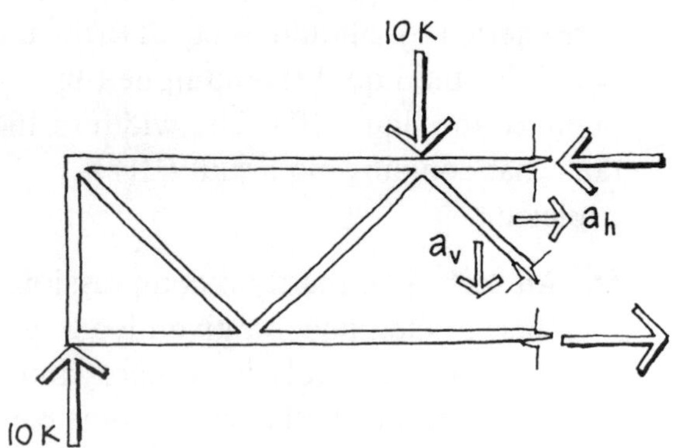

$$\Sigma V = 0$$

$$+10 - 10 - a_v = 0$$

$$a_v = 0.$$

As the vertical component of the force in member a is zero, the force in member a must be zero.

74.   B. A cable is essentially the reverse of an arch: an arch is stressed principally in compression, while a cable is stressed in tension. But where arches almost always have some bending moment combined with compression, cables are too flexible to have any significant resistance to bending.

75.   A. In a simple beam, where is the bending moment, and hence the bending stress, maximum? It varies with the loading, but it is usually close to midspan (B and D are incorrect). Within the depth of the beam, the bending stress is maximum at the outer fibers (A is correct) and minimum at the neutral axis (C is incorrect). These concepts apply to beams of any material, and the fact that the beam in this question is a glulam is immaterial.

76.   C. Let's look at the four choices. A is the loading diagram for a uniform load. B is the shear diagram, as it starts at zero and increases at a uniform rate; that is, the diagram is a straight line. As the slope of the moment diagram is equal to the value of the shear diagram, the slope varies from zero at the free end to a maximum value at the fixed end as shown in correct choice C. In choice D, the slope varies from maximum at the free end to zero at the fixed end, which is the reverse of what it should be.

77.   B. Questions on the exam sometimes test candidates' understanding of basic structural terms. The elastic limit (A) is the maximum unit stress that can be developed in a material without causing permanent deformation. The yield point (B) of a material is correctly defined in the question. The ultimate strength (C) is the maximum unit stress that can be developed in a material. The modulus of elasticity (D) is the constant ratio of the unit stress in a material to the accompanying unit strain, when the material is stressed below the elastic limit.

78. B. Beams whose reactions can be determined from the equations of equilibrium only are statically determinate. These include simple beams, cantilever beams (I), and overhanging beams with two supports (III). Beams whose reactions cannot be found from the equations of equilibrium only, but require additional equations, are called statically indeterminate beams. Continuous (IV) and fixed end beams (II) are statically indeterminate.

79. B. The best definition of redundancy is that in choice B. Long span structures generally have little redundancy; if overload or failure occurs, there is often no way for the system to redistribute loads to other parts of the structure. Consequently, there is greater potential for sudden failure of the entire structure. For seismic design, redundancy is often provided by secondary systems that can resist part of the lateral force if the primary system fails or is damaged.

80. B. We solve for R_2 by taking moments about R_1.

10 kips(4 ft.) + 10 kips(20 + 10) ft.
$\times R_2$(20 ft.) = 0
40 + 300 = 20R_2
R_2 = 340/20 = 17.0 kips
$\Sigma V = 0$
$R_1 + R_2 - 10 - 10 = 0$
$R_1 = 10 + 10 - R_2 = 10 + 10 - 17 = 3.0$ kips.

81. C. The section modulus of a beam (S), which is expressed in inches^3, is equal to the moment of inertia (I) divided by the distance from the neutral axis to the outermost fiber (c). Thus, S = I/c.

82. D. In a beam, the material close to the outermost fibers is highly stressed in flexure, while the material close to the neutral axis is understressed. By concentrating most of the beam's material towards the outer fibers and away from the neutral axis, as in a wide flange section, we obtain a very efficient bending member.

83. B. The first step is to calculate the value of the left reaction by taking moments about the right end.

R_L(20 ft.) − (1 kip/ft.)(20 ft.)(20 ft./2)
−10 kips (6 ft.) − 15 kips(16 ft.) = 0

$20R_L$ = 200 + 60 + 240

R_L = 500/20 = 25 kips

The shear at section $x - x = +25.0 - 15.0$ $-1(9) = 1$ kip, or 1,000#. It may be helpful to study the shear diagram for this beam, which is shown below.

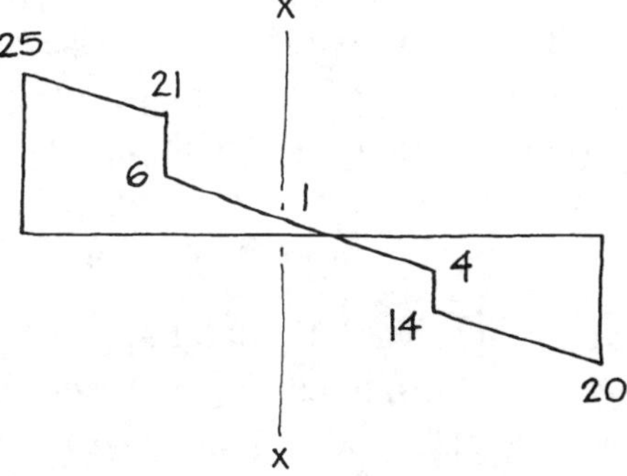

84. C. One way to solve this problem is to determine the loading on the beam. Let's look at the shear diagram again.

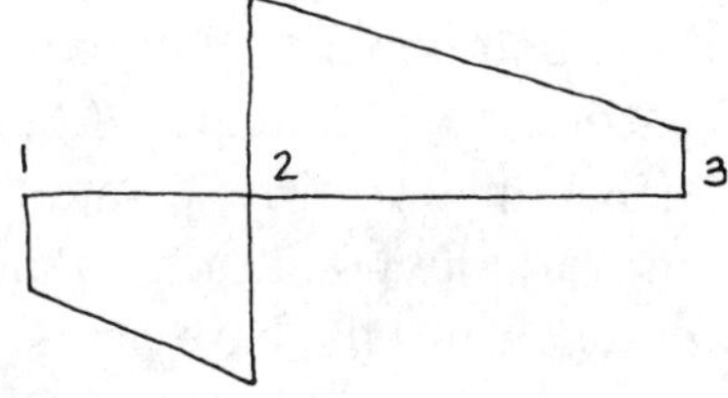

At point 1, the shear is negative, which corresponds to a downward vertical load. From 1 to 2, the shear diagram is a sloping line, which signifies a uniform load. At point 2, the shear changes from negative to positive, corresponding to an upward reaction. From 2 to 3, the shear diagram slopes the same as 1 to 2, and so we know that the uniform load is the same as from 1 to 2. At point 3, the shear is positive, which corresponds to a downward reaction. We can now draw the loading diagram.

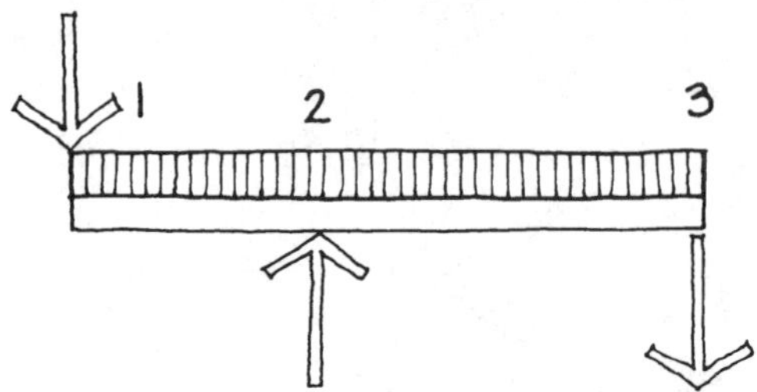

At 1, there is zero moment. From 1 to 2, there is negative moment, whose value increases to a maximum at support 2. From 2 to 3, the negative moment decreases and becomes zero at 3. C is therefore the correct moment diagram. Although D has the correct shape, it is incorrect because it shows positive moment in span 2–3.

85. D. Because the horizontal load acts to the right, the horizontal reactions must act to the left (C is incorrect). If we take moments about the left base, the horizontal load causes a clockwise moment. The vertical reaction at the right base must therefore cause a counterclockwise moment, and so the vertical reaction must act upward at the right base and downward at the left base (B is incorrect). That leaves A and D as possible answers. We isolate the left column from the point of zero moment (contraflexure) to the base and show the horizontal and vertical forces acting in the known directions, as shown below. *H* causes a clockwise moment about the base, and therefore *M* must be equal and counterclockwise, as shown in correct answer D. See the illustration in the top right-hand corner of this page.

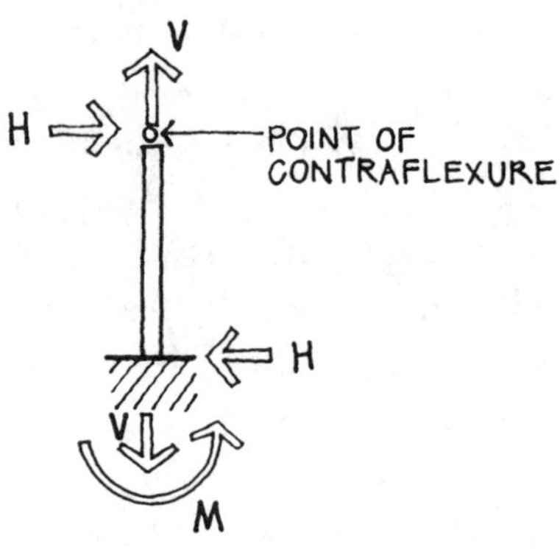

86. D. The floor framing system described in choice A (concrete fill, steel deck, composite beams, girders) is very widely used for a number of reasons, including economy, stiffness, simplicity, and ease of providing fire protection. The open web joist system (B) is also economical, but has less stiffness and is sometimes more difficult to fire protect than choice A. The concrete plank system described in C is often economical, quite stiff, and easily fire protected. The flat plate system (D), although economical in formwork, is not generally economical for spans longer than 25 feet.

87. A. The water in the tank exerts hydrostatic pressure against the tank wall, as shown in the horizontal section below.

This pressure tends to increase the diameter of the tank, or to lengthen its circumference. Any force that tends to stretch a member is tension. Another approach is to isolate one half of a horizontal section through the tank, as shown in the left column of page 85.

The net effect of the hydrostatic pressure is a horizontal force acting to the left. For

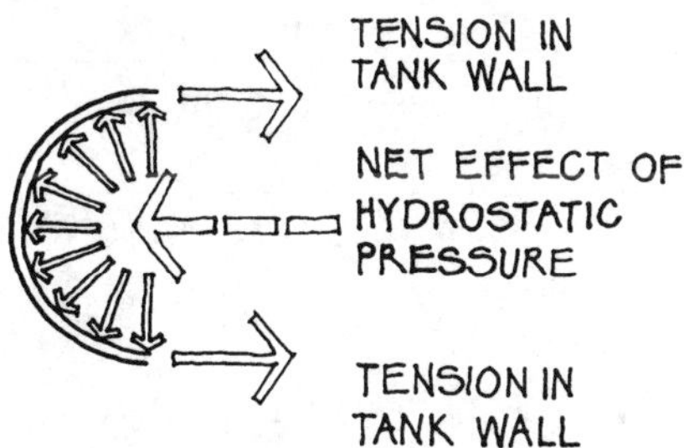

equilibrium, the internal force in the tank wall must act to the right, *away* from the cut section. A force that acts away from a section is tension.

88. **C.** The deflected shape of a beam is always a curve. B and D are composed of straight lines and are therefore incorrect. A shows the deflected shape for a simple beam. The beam in this question has fixed ends, which means that the ends are restrained against rotation, as shown in correct answer C.

89. **B.** A three-hinged arch is statically determinate; that is, its reactions can be computed using only the basic equations of equilibrium. The supports of a three-hinged arch permit rotation, while those of a fixed arch do not. Fixed arches are therefore stiffer than hinged arches (A is incorrect). Three-hinged arches are able to resist horizontal loads (C is incorrect), and can have a variety of profiles, not limited to semicircular shapes (D is incorrect).

90. **C.** The structural divisions of the exam invariably include a number of questions on architectural history, such as this one. One of the features of Byzantine architecture was the dome, which, during the later Renaissance period, was sometimes raised on a drum and surmounted by a lantern. The transition from the round dome or drum to the square base was effected by spherical pendentives, triangular in form, which rested on pillars at the four corners of the space below. See the illustration in the top right-hand corner of this page.

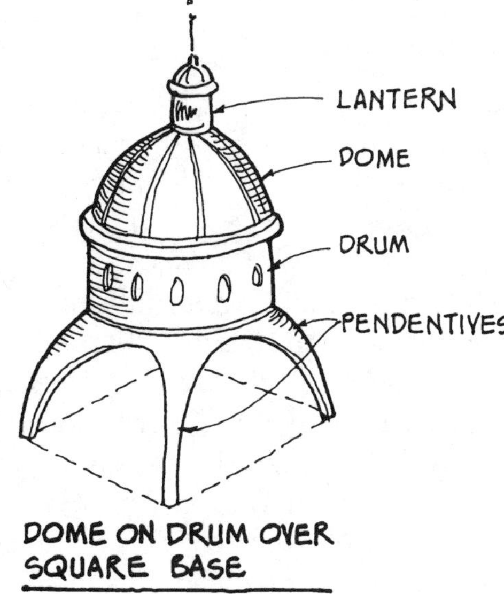

DOME ON DRUM OVER SQUARE BASE

91. **D.** The campanile of the Cathedral of Pisa, better known as the Leaning Tower of Pisa, is probably the world's most famous example of differential settlement (II). The tilting began during the tower's construction (I) and continues to this day; its top is now out of plumb by 16 feet, not 16 inches (III is incorrect). The tilt of the tower cannot be blamed on its structure; it is an elegant piece of architecture, consisting of a cylindrical core surrounded by marble columns that form a balanced, symmetrical design.

92. **B.** Pier Luigi Nervi was one of the giants of 20th-century architecture and engineering. Although trained as an engineer, his aesthetic sense and design vision led him to design structures of exceptional beauty. He was the first designer to successfully use ferrocement (A), a combination of steel mesh and cement mortar. His concrete lamella roofs (C), such as the airplane hangar roofs he built during the 1940s, were among the largest and most elegant roofs of this kind ever built. For the Borgo paper plant, Nervi designed a roof that hung from the cables of a suspension bridge (D). However, Nervi is not associated with the tubular frame (B), a concept used by Fazlur

Khan and others in the design of very tall buildings.

93. A. Questions involving shear and moment diagrams often appear on the exam. Among the concepts that should be understood are: (1) the slope of the moment diagram is equal to the value of the shear at that point and (2) the bending moment is maximum where the shear passes through zero. In this case, the slope of the moment diagram is constant and positive between points 1 and 2, and therefore the shear is constant and positive between 1 and 2. From 2 to 3, the slope of the moment diagram is also constant and positive, but less than the slope between 1 and 2. Therefore, the shear has a constant positive value between 2 and 3, which is less than that between 1 and 2. At point 3, the moment is maximum, and therefore the shear passes through zero at point 3. Between 3 and 4, the slope of the moment diagram is constant and negative, and therefore the shear is constant and negative between 3 and 4. The only shear diagram that meets all these criteria is shown in correct choice A.

94. B. If a deflection question should appear on the exam, it is likely that the necessary deflection formula will be provided. In this case, we are given the formula $\Delta = 5wL^4/384EI$. As $wL = W$, the total load on the beam, the formula becomes $\Delta = 5WL^3/384EI$.

$\Delta = 1.0$ inch

$W = 1,500\#/\text{ft.} \times 28 \text{ feet} = 42,000\#$

$L^3 = (28 \text{ feet} \times 12 \text{ in./ft.})^3$

$\quad = 37,933,056 \text{ in.}^3$

$E = 29,000,000 \text{ psi}$

Thus

$$1.0 = \frac{5(42,000)(37,933,056)}{384(29,000,000)(I)}$$

$$I = \frac{5(42,000)(37,933,056)}{384(29,000,000)(1.0)}$$

$$= 715.3 \text{ in.}^4 \text{ (answer B)}$$

95. C. To solve this problem, we resolve the 1,500# force acting at 45° with the horizontal into its horizontal and vertical components. Its horizontal component = 1,500 cos 45° = 1,500(.707) = 1,060.5# to the right, and its vertical component = 1,500 sin 45° = 1,500(.707) = 1,060.5# upward. The sum of the horizontal forces is 1,500# + 1,060.5# = 2,560.5# to the right. The only vertical force is 1,060.5# upward. The resultant is $\sqrt{2,560.5^2 + 1,060.5^2} = 2,771\#$. Tan θ = 1,060.5/2,560.5 = 0.414, from which θ = 22.5°, as shown below. C is therefore the correct answer.

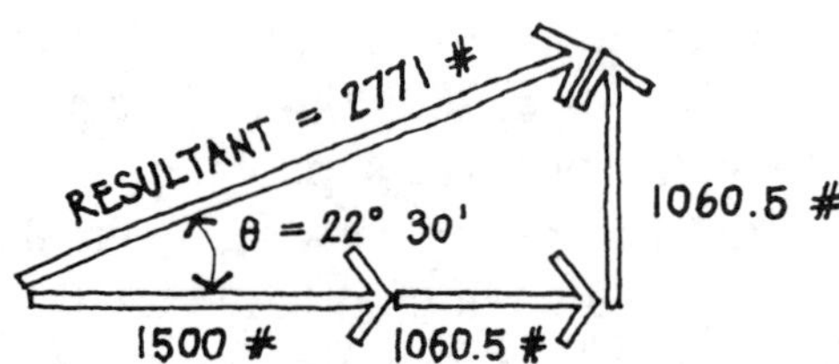

96. A. The best definition of live load is that given in correct answer A. B is incorrect because a continuously applied load is usually considered to be dead load. C is incorrect because live load does not include wind or earthquake load. D is not as inclusive as A, and is therefore incorrect, because live load may include loads other than occupants and movable furniture.

97. A. In this question, we are not asking which building type has the greatest structural cost, but for which building type is the structural cost the lowest percentage of the total cost of construction. As a rule of thumb, the structural cost for commercial or office buildings averages about 25 percent of the total cost of construction (D is incorrect). For buildings with minimal architectural and mechanical requirements, such as warehouses and parking garages, the structural cost may be 50 percent or more of the total cost of construction (B and C are incorrect). For buildings that have complex or expensive architectural and mechanical requirements, such as hospitals, the structural cost may be only 10 or 15 percent of the total cost of construction. The correct answer is therefore A.

98. **B.** A structural system is an arrangement of structural components that resists a building's vertical and/or horizontal loads. There are many factors that influence the choice of structural system, including the building's spans and loads. This question tests your understanding of the appropriateness of various structural systems for the given conditions. The one-way concrete joist and beam system (I) is generally economical for spans of 35 to 40 feet, and would therefore not be economical in this case. Similarly, the flat slab system (III) is appropriate for spans of about 25 to 30 feet, and would not be economical. Prestressed concrete systems, such as those in choices II and IV, are usually able to span longer distances. The correct answer is therefore B.

99. **C.** The exam often includes a few problems in statics, similar to this one, which can be solved by using the basic equations of static equilibrium: $\Sigma H = 0$, $\Sigma V = 0$, $\Sigma M = 0$. In this case, there is a pin at B, which means that there are vertical and horizontal reactions there, but no moment. At A, the horizontal bar pinned at both ends can resist an axial horizontal reaction only, but no vertical reaction or moment. We take moments about B to solve for A_H. See the illustration in the top right-hand corner of this page.

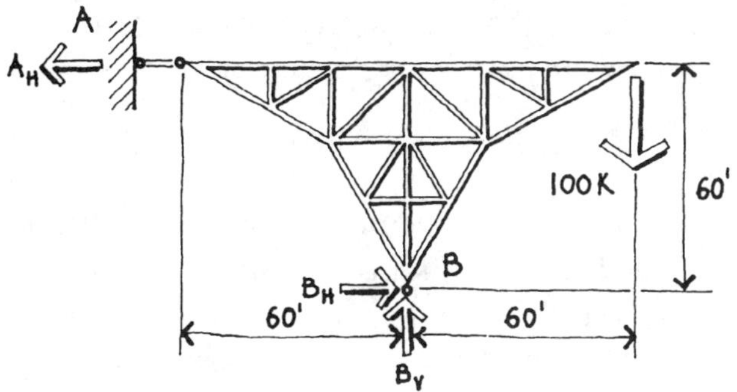

$\Sigma M_B = 0$

$+ 100 \text{ kips}(60 \text{ ft.}) - A_H(60 \text{ ft.}) = 0$

$A_H = 100 \text{ kips}$

$-A_H + B_H = 0$

$B_H = A_H = 100 \text{ kips}$

$\Sigma V = 0$

$-100 \text{ kips} + B_V = 0$

$B_V = 100 \text{ kips}$

The reactions are shown correctly in choice C.

100. **B.** The exam includes both numerical and conceptual questions, although there are usually more of the latter. This question tests your understanding of the concept of deflection. All beam deflection formulas are essentially in the form $\Sigma = KL^3/EI$, where K is a constant that depends on the load and the loading condition. Thus, to reduce the deflection, one must increase the moment of inertia I (correct answer B). A and D are incorrect because all steel, regardless of yield strength, has the same modulus of elasticity E. C is also incorrect: while a beam with a greater section modulus has a greater ability to resist flexural stress, it does not necessarily have a greater value of moment of inertia I.

101. **A.** Statements I and II are correct. Statement III is incorrect; the horizontal thrust at each end of a cable is inversely proportional to the sag of the cable. Thus, the

greater the sag, the smaller the thrust. A is therefore the correct answer.

102. C. Hooke's Law states that up to a certain unit stress, called the elastic limit, unit stress is directly proportional to unit strain. This constant ratio of unit stress to unit strain for a given material is called the modulus of elasticity (C) and is represented by the letter E.

103. A. Candidates should understand basic structural terminology. The stiffness of a member refers to its resistance to deformation (correct answer A). For a member with axial load, the stiffness is a function of the modulus of elasticity E of the material and the cross-sectional area A of the member. For a flexural member, the stiffness refers to its resistance to deflection and depends on E and on the moment of inertia I of the member.

104. B. Simple beams differ from continuous beams in a number of ways: the maximum positive moment in the simple beam is greater (A), the maximum deflection of the simple beam is greater (C), and the continuous beam has negative moment over its interior supports, while the simple beam never has any negative moment (D). Thus, A, C, and D are all correct statements. B is the incorrect statement we are looking for: the maximum shear in the simple beam is wL/2, which is less than that in the continuous beam, which is 5wL/8.

105. D. Candidates should have some familiarity with typical stress-strain diagrams, such as that shown in the right column, for steel in tension. The elastic limit (correct answer D) is defined as the maximum unit stress that can be developed in a material without causing permanent deformation when the stress is released. The *modulus of elasticity* (A) is the ratio of unit stress to unit strain

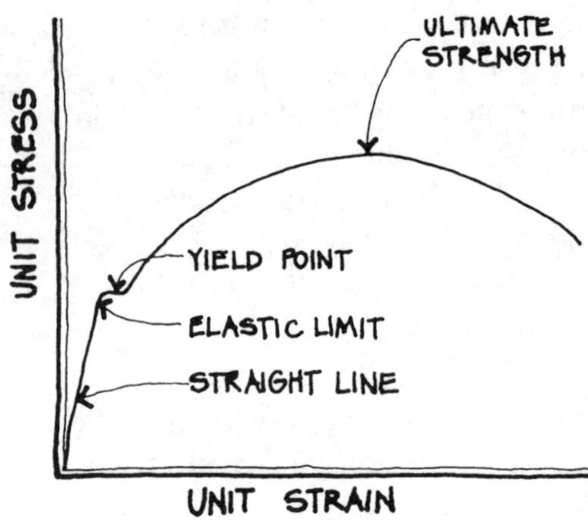

for the straight-line portion of the diagram, when the unit stress is below the elastic limit. The *ultimate strength* (B) is the maximum unit stress that can be developed in a material, and the *yield point* (C) is defined as the unit stress at which a material continues to deform without an increase in load.

106. A. Questions testing your understanding of structural behavior often appear on the exam. Structural steel columns fail by buckling. The ratio Kl/r is a measure of the buckling tendency of a steel column; the larger the value of Kl/r, the greater the tendency of the column to buckle, resulting in a lower column capacity. In this ratio, K is a constant determined by the end conditions of the column (I), that is, whether the column ends are pinned or fixed and whether the column is free to translate (move laterally). 1 is the actual unbraced length of the column and K_l is the effective length (II is incorrect). The radius of gyration r is a property of the column cross-section and is independent of the yield strength of the steel (III is incorrect). Because only I is a correct statement, A is the correct answer.

107. B. Handrail problems, such as this one, sometimes appear on the exam. Such problems are easily solvable using the basic principles of statics. In this problem, we solve for the force in bolt A by taking moments about bolt B.

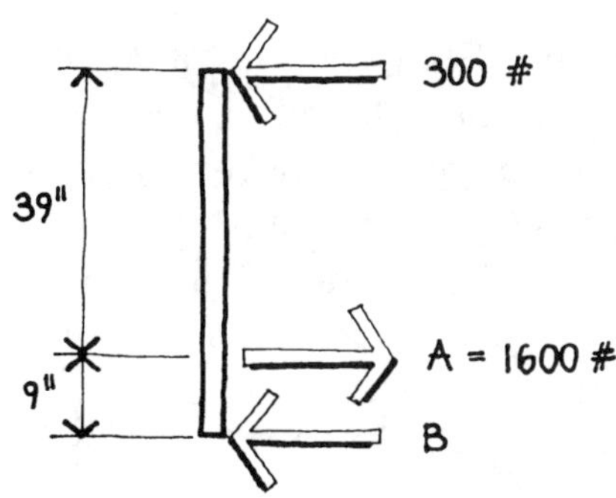

$$\Sigma M_B = 0$$

$$-300\#(39" + 9") + A(9") = 0$$

$$A = (300 \times 48) \div 9 = 1{,}600\#$$

Bolt A is in *tension*, because the force acts *away* from the vertical post (correct answer B).

108. C. You may believe that this is such a simple, basic problem that there must be a trick to it. No, there is no trick; this is just a straightforward problem, like most exam problems. The unit stress in the bar is simply P/A, where P is the load of 50,000 pounds and A is the cross-sectional area of the two-inch-diameter bar = $\pi(2)^2/4$ = 3.14 square inches. So the unit stress is equal to 50,000 lbs. ÷ 3.14 square inches = 15,915 psi (answer C). The length of the bar is irrelevant.

109. A. A statically determinate beam is one whose reactions can be determined by using the basic equations of static equilibrium ($\Sigma H = 0$, $\Sigma V = 0$, $\Sigma M = 0$). Examples include simple beams and overhanging beams on two supports (IV). Beams whose reactions cannot be determined from the equations of equilibrium only, but require

additional equations, are called statically indeterminate beams. These include beams fixed at both ends (I); beams fixed at one end and simply supported at the other end (II); and continuous beams (III). A is the correct answer.

110. C. Each radial cable supports vertical load and resists tension, as shown.

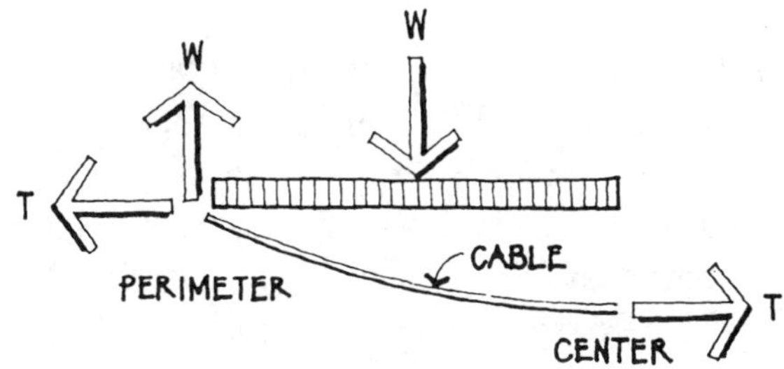

For equilibrium, the center ring must be *lower* than the perimeter ring (A is incorrect). Each cable *pulls* on the center ring, which tends to expand it, thus stressing it in tension. Each cable also *pulls* on the perimeter ring, which compresses it (C is correct, B and D are incorrect).

111. D. Candidates should be able to determine the reactions of any statically determinate beam subject to any load. To determine the reaction at R_1, we take moments about R_2.

$$\Sigma M_{R_2} = 0$$

$$+(5{,}000\# \times 8\text{ ft.}) - (5{,}000\# \times 16\text{ ft.})$$

$$-\left(1{,}000\#/\text{ft.} \times 24\text{ ft.} \times \frac{24\text{ft}}{2}\right) + (R_1 \times 24\text{ ft.})$$

$$= 0$$

$$R_1 = (-40{,}000 + 80{,}000 + 288{,}000) \div 24$$

$$= 328{,}000 \div 24$$

$$= 13{,}667\# \text{ (answer D)}.$$

112. C. The modulus of elasticity E of a material is defined as the ratio of unit stress to unit strain when the unit stress is below the elastic limit. Because unit stress = P/A and unit strain = Δ/L,

$$E = (P/A) \div (\Delta/L) = PL/A\Delta$$

$$= (50{,}000\# \times 10 \text{ ft.} \times 12 \text{ in./ft.})$$

$$\div \left[\frac{\pi(2\text{in})^2}{4} \times 0.159\text{in} \right]$$

$$= 6{,}000{,}000 \div 0.4995 = 12{,}012{,}000 \text{ psi}$$

(answer C).

113. D. The nature of long span construction requires a comprehensive quality control program, including proper field inspection and testing (I). Snow drifts, partial snow loads, and the effects of wind and earthquake demand special attention in long span structures (II). The effects of temperature, creep, and shrinkage are more pronounced in long span structures (III). And long span structures are sensitive to secondary stresses caused by deflection and the interaction of building elements (IV). Because all four factors are more critical for long span buildings than for conventional buildings, D is the correct answer.

114. A. The moment over the interior supports of a continuous beam is always negative when the beam supports downward loads (B is correct). We isolate span 1-2.

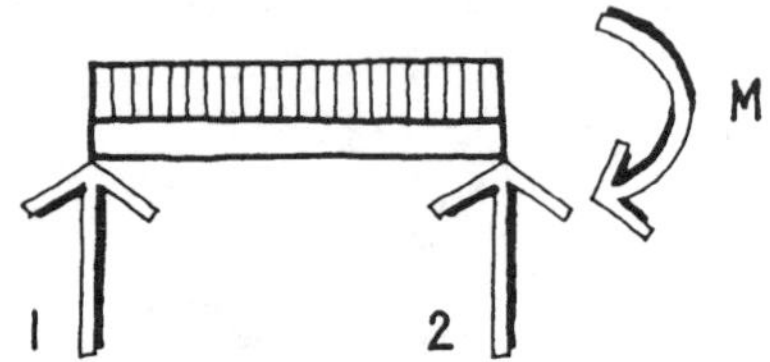

The moment at 1 is zero, the moment at 2 is negative (tension at the top, compression at the bottom), and the moment diagram is parabolic because the load is uniform. The moment diagram for span 1-2 is therefore as shown.

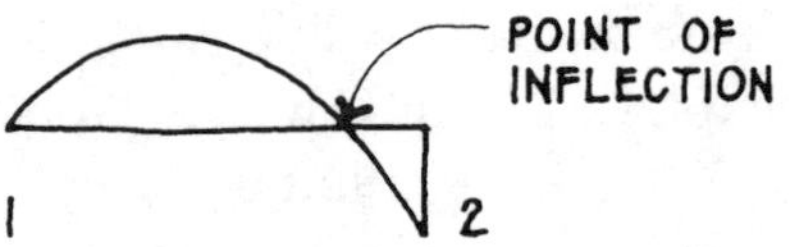

There is positive moment from 1 to the point of inflection and negative moment from there to 2. A is therefore the incorrect statement we are looking for. We now isolate span 2-3 and take moments about 2, assuming that the reaction at 3 is downward.

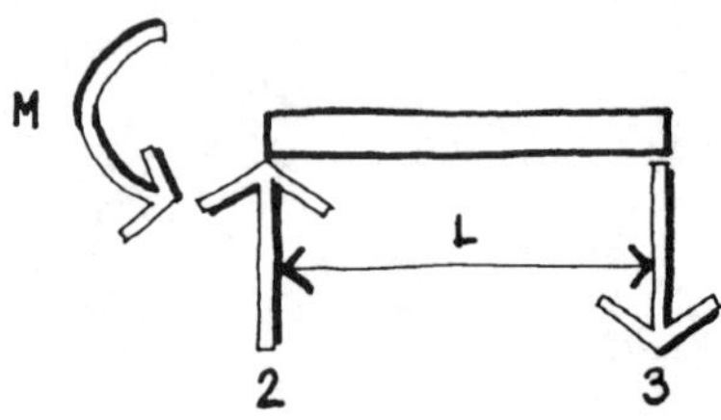

$$\sum M_2 = 0$$

$$-M + (R_3 \times L) = 0$$

$$R_3 = +M/L$$

Because R_3 comes out positive, our assumption that it acts downward is correct (C is correct). The moment in span 2-3 is always negative and varies from zero at 3 to M at 2 (D is correct).

115. B. One of the most fundamental concepts in structures is that of equilibrium. For an object to be in equilibrium, it must have no unbalanced force acting on it (A and D are correct). It must also have no unbalanced moment acting on it (C is correct). B is incorrect, and is therefore the answer to this question, because if there is a resultant force on an object, it will not be in equilibrium.

116. B. A membrane can only resist tension. The vertical component of the tension in the membrane is an uplift force (III), while the horizontal component compresses the perimeter ring (I), making B the correct answer.

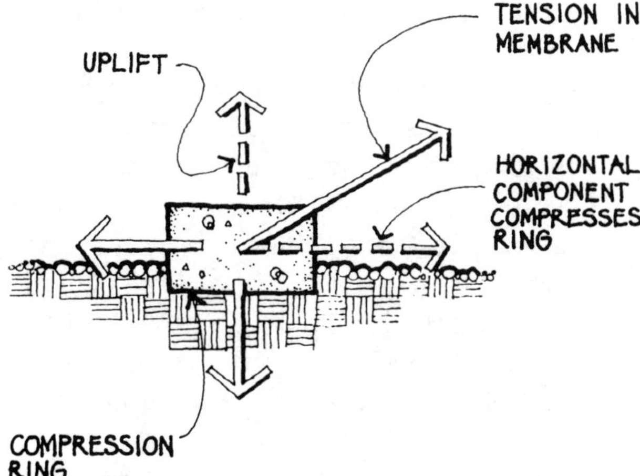

117. B. The pressure exerted by a liquid is equal to the unit weight of the liquid multiplied by the depth of the liquid. At the top of the tank, the pressure is 62.4#/cu.ft. $\times$ 0 ft. = 0. At one foot down, the pressure is 62.4 $\times$ 1 = 62.4#/sq.ft., at two feet down, the pressure is 62.4 $\times$ 2 = 124.8#/sq.ft., and so on. At the bottom of the tank, the pressure is 62.4 $\times$ 10 = 624#/sq.ft. Thus, the pressure varies linearly, from zero at the top to 624#/sq.ft. at the bottom, as shown correctly in answer B.

118. D. In this problem, we must determine the directions of the reactions but not their magnitudes. We draw a free body diagram and assume the reactions act in the directions shown.

$$\Sigma M_2 = 0$$

$$+ (P \times L) - (1_H \times b) = 0$$

$$1_H = +PL/b$$

Because 1_H comes out positive, our assumption that it acts to the left is correct. From $\Sigma H = 0$, $2_H = 1_H$ and acts to the right. We next draw a free body diagram of horizontal member 2-3-4.

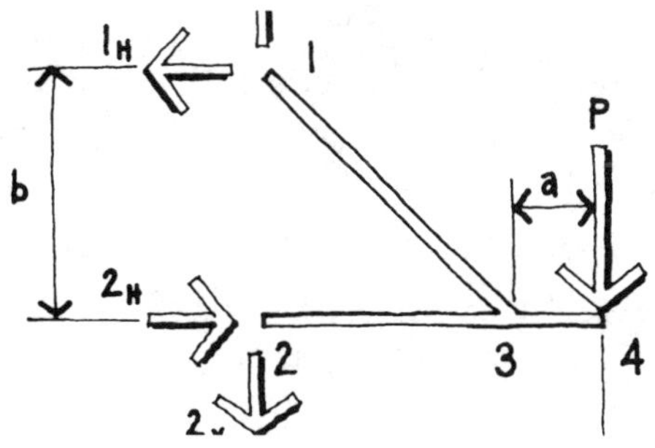

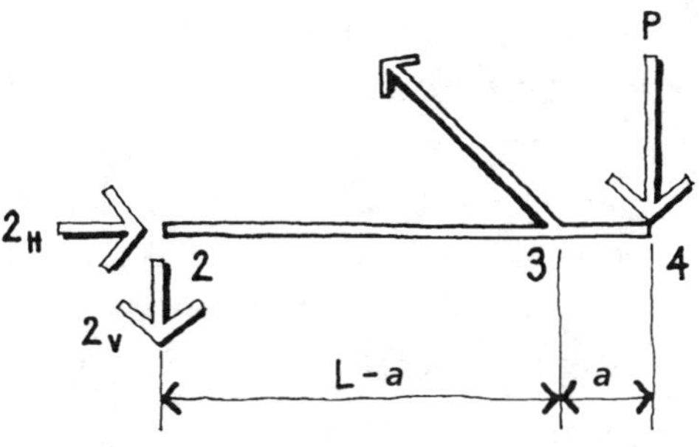

$$\Sigma M_3 = 0$$

$$+(P \times a) - 2_v \times (L - a) = 0$$

$$2_v = +Pa/(L - a)$$

Because 2_v comes out positive, our assumption that it acts downward is correct.

$$\Sigma V = 0$$

$$+1_v - 2_v - P = 0$$

$$1_v = 2_v + P$$

Because 1_v comes out positive, our assumption that it acts upward is correct. The directions of the reactions are shown correctly in answer D.

119. B. In answering this question, we first review a few basic definitions and formulas. Unit stress is equal to axial load P divided by cross-sectional area A, or P/A. Unit strain is equal to total strain Δ divided by length L, or Δ/L. Modulus of elasticity E is equal to unit stress divided by unit strain, or $E = (P/A) \div (\Delta/L) = PL/A\Delta$.

Transposing, $\Delta = PL/AE$. Thus, the change of length Δ is directly proportional to the load P (I) and the length L (II) and inversely proportional to the cross-sectional area A (IV) and the modulus of elasticity E (V). The moment of inertia (III) does not affect the change of length. B is therefore the correct answer.

120. B. The flexural stress in a rectangular beam varies from a maximum value at the outer fibers to zero at the neutral axis, as shown in answer A. The shear stress varies parabolically, from zero at the outer fibers to a maximum value at the neutral axis, as shown in correct answer B. Answers C and D are arbitrary diagrams.

121. D. A three-hinged gabled frame has a hinge at the center and at each support and is statically determinate (A is correct). The hinges at the supports permit rotation (B is correct), and the hinge at the center also permits rotation, which prevents any moment from being developed at the center (C is correct). D is the incorrect statement and is therefore the answer to this question: the maximum moment does occur at the intersection of the column and the sloping beam, but this moment is generally less than in a rectangular rigid frame.

122. A. A parallel chord truss is analogous to a steel beam: the truss chords are like the beam's flanges, while the truss web members are similar to the beam's web. Just as the beam flange forces increase toward the center of the span, the forces in the truss chords increase toward the center of the span. And just as the shear in the beam web decreases toward the center of the span, similarly the forces in the truss web members decrease toward the center of the span. A is therefore the correct answer.

123. D. Exam questions sometimes test candidates' understanding of the economics of various structural systems and components. This is that kind of question. The connections used in structural steel systems comprise a significant part of the cost of these systems, and can even influence the type of structural steel system selected (A is correct). It is true that fillet welds are usually more economical than full penetration welds (B is correct). Shop connections are usually preferred over field connections because they are less costly (C is correct). In addition, good quality control is easier to achieve in the shop than in the field. Depending on the circumstances, welded connections are not necessarily more economical than bolted connections. D is therefore the incorrect statement.

124. B. This can be a troublesome question because the choices are similar, but have subtle differences. When a rigid frame supports a uniform vertical load, the horizontal member deflects downward, as in choices A, B, and C. D is the shape of the deflected frame when it resists a lateral load, such as wind or earthquake, and is therefore incorrect. C is the deflected shape of a simple beam supported by columns, not a rigid frame, and is therefore incorrect. A is the deflected shape of a rigid frame with hinged bases, not fixed (note that the column bases are rotated, not vertical). B is similar to A, but note that the bottoms of the columns are vertical where they meet the ground, which indicates a fixed condition. B is therefore the correct choice.

125. C. Each force produces a moment about point O equal to the magnitude of the force multiplied by its distance from point O. The total moment is equal to the algebraic sum of the moments caused by all the forces.

The 1,000# horizontal force produces a clockwise moment equal to 1,000# × 4 feet = 4,000 ft.-lbs. The 1,000# vertical force produces no moment, because its line of action passes through point O. The moment caused by the 1,000# force at 30°C is determined by resolving the force into its vertical and horizontal components. The vertical component is 1,000 sin 30°C = 500#, and the horizontal component is 1,000 cos 30°C = 866#. The vertical component causes no moment, because its line of action passes through point O. The horizontal component produces a clockwise moment of 866# × 4 feet = 3,464 ft.-lbs. The total moment is the algebraic sum of the moments = 4,000 + 3,464 = 7,464 ft.-lbs. (answer C).

126. C. Steel columns rest on and are generally welded to steel base plates, which transfer the column load by bearing on the concrete foundation. The bearing pressure under the base plate is equal to P/A, where P is the column load and A is the area of the base plate = 300,000# ÷ (20 in. × 20 in.) = 750 psi (correct answer C). This is as simple a calculation as you're likely to see on the exam.

127. A. If a parabolic arch supports a uniformly distributed vertical load, it will be stressed only in compression, with no bending (A is correct). Under any other loading, it will be subject to some bending moment (C). B (pure bending) describes the stress in a beam, not an arch, and D (pure tension) describes the stress in a cable, not an arch.

128. B. The dead load of a structure acts continuously, but the live load can vary from zero to the full design load. In the design of the structural framing of a building, the live load must be arranged so as to produce the maximum moments. For a simple beam, the full design live load acting on the entire span will produce the maximum positive moment. For a continuous beam, the live load arrangements which will produce the maximum moments are as follows: (1) live load on two adjacent spans will produce the maximum negative moment over the support, and (2) live load on alternate spans will produce the maximum positive moment between supports. In this case, therefore, arrangement C (live load on spans 1-2 and 2-3) will produce the maximum negative moment over support 2, while arrangement B (live load on spans 1-2 and 3-4) will produce the maximum positive moment in spans 1-2 and 3-4. Arrangements A and D will not produce either maximum positive or negative moment. B is therefore the correct answer.

129. B. The maximum stress that a steel column can resist without buckling is a function of its slenderness ratio K1/r, where K is a factor that depends on the column end conditions, 1 is the unbraced length of the column, and r is the radius of gyration of the column section, which is equal to $\sqrt{I/A}$. I is the moment of inertia of the column section (choice II) and A is the cross-sectional area of the column section (choice III). Therefore, B is the correct answer.

130. D. Candidates should have some knowledge concerning the efficiency, economy, and appropriateness of various structural systems and components. In general, we try to keep everything as simple, regular, and conventional as possible. For example, as moment or rigid connections are more expensive than simple or shear connections, the number of moment connections should be kept to a minimum (I is incorrect). Rolled sections should be used wherever possible; while the use of built-up sections may save weight, the additional labor cost often results in a higher in-place cost (II is incorrect). III is also incorrect; high-strength steel is stronger than ASTM A36 steel, but it is also more costly and therefore not always the more economical choice. IV is the only correct statement (D is correct). Most steel floor decking today is composite because it is more economical than non-composite decking.

131. D. In the post-and-beam, the maximum moment in the beam is equal to $wL^2/8$ and there is no moment in the columns. In the rigid frame, the beam ends are restrained by the columns, which reduces the maximum moment in the beam (II) and results in moment in the columns (I). The axial force in the columns is the same for both the rigid frame and the post-and-beam (III). The column bases of the rigid frame have a horizontal reaction, which the column bases of the post-and-beam do not (IV). Because all four statements are correct, the answer is D.

132. A. All structures tend to expand when the temperature increases and contract when the temperature decreases. As the bottom of a dome is prevented by its foundation from expanding and moving outward when the temperature rises, the only way the dome can expand is to move upward (correct answer A).

133. C. Studies have shown that the optimum bay shape for steel framing is rectangular with a length-to-width ratio of about 1.3 (A and D are incorrect). The beams should preferably span in the long direction and the girders in the short direction (C is correct, B is incorrect).

134. D. A frame with diagonal braces (A) is a braced frame, not a rigid frame. A frame that supports a building's vertical loads (B) may or may not be a rigid frame. A rigid frame may have fixed bases (C), but it may also have hinged bases. The best, most inclusive answer is D: a rigid frame is a frame with rigid joints capable of resisting moment.

135. A. One of the primary considerations in hospital design is providing maximum flexibility without interruption of service. In this regard, interstitial trusses about eight feet in depth (A) offer unusual layout flexibility by providing a walk-through ceiling space above each column-free hospital floor. This interstitial space is utilized for all mechanical and electrical services, which can be maintained or changed without disrupting normal hospital functions on the floor above or below.

136. A. The traditional method for structural steel design is Allowable Stress Design (ASD), in which the actual dead and live loads are unfactored, that is, not increased. The factor of safety is obtained by using allowable stresses that are less than the yield stress. For example, the allowable flexural stress is 0.66 times the yield stress, which results in a factor of safety of 1Π 0.66, or 1.5. Therefore, I is correct. Load and Resistance Factor Design, sometimes called limit states design, is a newer method of designing steel structures, in which the various loads (dead, live, etc.) are multiplied by their respective load factors. The nominal strength, which is most often the yield strength, is multiplied by a resistance factor Ø, which is generally less than 1. Thus, II is correct. The primary objective of LRFD is to provide more uniform reliability for all steel structures under various loading conditions (III is incorrect). A is therefore the correct answer.

137. B. In this question, the four choices are all well-known Chicago buildings of the late-19th century. Burnham and Root's 16-story Monadnock Building (A) was the last of the pure masonry towers, with six-foot-thick walls at its base. Louis Sullivan's Carson Pirie Scott Store (C), with its metal frame, broad windows, and unique decoration, was the most modern building of the Chicago School. For the Marshall Field Warehouse (D), H. H. Richardson designed exterior masonry piers and arches with interior framing of wood and iron. But it was Jenney's Home Insurance Building (correct answer B) that is considered the first iron and steel-framed skyscraper. Although only ten stories high, it was the predecessor of all the tall metal-framed buildings that followed.

138. A. Three of the buildings in this question are essentially hollow steel tubes that cantilever from their foundations when subject to lateral wind or earthquake forces. The Sears Tower in Chicago (C) consists of nine tubes that end at varying heights. The John Hancock Building in Chicago (D) and the Bank of China Tower are both gigantic trussed tubes. Only the Water Tower Building in Chicago (correct answer A) is not a tubular steel building, but rather a very tall reinforced concrete skyscraper.

139. D. Pier Luigi Nervi designed buildings that expressed their structure in bold and imaginative ways. His work is associated with all four structural concepts listed (correct answer D). He made use of ferrocement (I), a material consisting of layers of wire mesh embedded in concrete mortar. The Palazzeto Dello Sport has a prefabricated ribbed concrete shell dome (II), and

the airplane hangars he built for the Italian Air Force in the 1930s had prefabricated concrete lamella roofs (III). Nervi's Borgo Paper Plant utilized a suspension bridge to achieve a span of 830 feet.

140. B. If the only load acting on a cable is its own weight, the shape the cable assumes is a catenary (correct answer B), similar to a parabola. If the loads were uniformly distributed horizontally across the span, the cable would assume the shape of a parabola (incorrect answer A). C and D are also incorrect.

141. C. A thin shell is a structure with a curved surface that supports load by compression, shear, and tension in its own plane (I, II, and III), but is too thin to resist any bending stresses (IV is incorrect). C is therefore the correct answer. Among the more popular thin shell shapes are the dome, the cylindrical or barrel shell, the vault, and the hyperbolic paraboloid.

142. D. The section modulus of a beam (S) is equal to the moment of inertia of the beam (I) divided by the distance from the outermost fiber of the beam to the neutral axis (c), or $S = I/c$. The maximum bending stress $f = M/S$, so the greater the section modulus, the lower the bending stress for a given moment. Therefore, the section modulus is a measure of the beam's bending strength (D is correct).

143. B. Candidates should be aware that this kind of question has appeared on the exams before, and therefore might appear again. According to the Building Code Requirements for Reinforced Concrete (ACI 318), which has been incorporated into many building codes, the minimum concrete coverages are 3" for footings cast against earth (correct answer B).

144. C. The answer is *I, II, and IV.* Live load is the load superimposed by the use and occupancy of the building, such as furniture (III), and does not include wind load, snow load, earthquake load, or dead load. (I, II, and IV are not live load.)

145. B. The answer is 134 kips. To solve this problem, we use Tables 1-D and 1-E from the AISC Manual of Steel Construction (see pages 62 and 63). From Table 1-D, the allowable load in double shear for each 7/8" bolt conforming to ASTM A325-X (threads excluded from shear plane) = 36.1 kips. From Table 1-E, the allowable load in bearing for each 7/8 bolt = 60.9 kips per inch of thickness of material. The wet of the W24 × 76 beam on which the bolts bear has a thickness of 0.440". Each bolt therefore has an allowable load in bearing of 60.9 × 0.440 = 26.8 kips. This is the governing value, as it is less than the allowable load in shear of 36.1 kips. The allowable load of the connection is therefore 26.8 kips per bolt times 5 bolts, or 134 kips.

146. A. When the use or occupancy of a building changes, the building must comply with all building code regulations for the new use, including live load requirements (A). The new use is not required to be less hazardous than the existing use (B is incorrect); in fact, it can even be more hazardous, provided that the building complies with the code regulations for the greater hazard. C is obviously incorrect, as an architect has no proprietary interest in a building after its completion. D is also incorrect; the structural design of the building would be checked only if the new use has more severe requirements than the existing use.

147. D. Even if you are not familiar with specific code requirements, it should be obvious that a factor of safety against sliding should be provided. The only answer that has a factor of safety is D.

148. A. Although an arch is often subject to some bending moment, its internal stresses are primarily compression.

149. A. The allowable stresses for wood members and fasteners tabulated in building codes apply to normal duration of loading. For shorter durations of load, the allowable stresses may be increased; the shorter the duration, the greater the allowable stresses (correct answer A). For example, for two months' duration, as for snow, the allowable stresses may be increased 15 percent.

150. A. Most building codes require the dead load resisting moment to be at least 1.5 times the overturning moment caused by earth pressure, making A the correct answer. If the dead load resisting moment were less than 1.5 times the overturning moment, the simplest solution would be to make the footing wider. Making the footing deeper would increase the dead load resisting moment slightly, and increasing the amount of reinforcing steel would have no effect on the dead load resisting moment.

151. A. The traditional method for structural steel design is Allowable Stress Design (ASD). In this method, the actual dead, live, and other loads are unfactored, that is, not increased. The factor of safety is obtained by using allowable stresses that are less than the yield stress. For example, the allowable flexural stress is 0.66 times the yield stress, which results in a factor of safety of 1 ÷ 0.66, or 1.5. Statements III and IV therefore apply to Allowable Stress Design. Load and Resistance Factor Design, sometimes called limit states design, is a different and newer method of designing steel structures. In this method, which is similar to the strength design method for reinforced concrete design, the various loads (dead, live, etc.) are multiplied by their respective load factors (statement I). For example, the load factor for dead load only is 1.4 while the load factors for a member supporting dead, live, and roof live load are, respectively, 1.2, 1.6, and 0.5. The nominal strength, which is most often the yield strength, is multiplied by a resistance factor ϕ, which is generally less than 1 (statement II). For example, ϕ for flexure is equal to 0.90 and ϕ for compression is equal to 0.85. The primary objective of LRFD is to provide more uniform reliability for all steel structures under various loading conditions. As of this writing, the exam is likely to have only a few questions involving LRFD. However, as this method gains more widespread acceptance, we can expect the exam to include a greater number of questions on the subject.

152. B. Wide flange sections are commonly used for beams, girders, and columns. They are efficient bending members because most of their material is in the flanges, at the greatest possible distance from the neutral axis. Looking at the other answers, wide flange beams have relatively thin webs (A is incorrect), their strength is much greater about the horizontal axis than the vertical axis (D is incorrect), and although they are symmetrical about both axes, that is unrelated to their efficiency as bending members (C is incorrect).

153. C. The footing shown is called a cantilever footing or strap footing. It consists of an exterior column footing joined by a concrete beam to an interior column footing and is used where the exterior column is too close to the property line to have a symmetrical, concentric footing. A combined footing (B) is similar, except that it is one footing supporting both columns, not two separate, connected footings. A mat foundation (A) is essentially one large footing under the entire building area, and a pile cap (D) is a reinforced concrete element used to transmit a column load to a group of piles.

154. D. The basic structural component in all suspension structures is a steel cable, which has great tensile strength but practically no compressive or bending strength.

155. A. The capacity of a wood column is determined by several factors: the modulus of elasticity (E) of the wood, which in turn depends on its species and grade; the allowable compressive stress (F_c), which also depends on the species and grade of the wood; the ratio $1/d$, where 1 is the unbraced height of the column and d is the least lateral dimension of the column, and the cross-sectional area of the column. Therefore B, C, and D are all correct. The radius of gyration is used in determining the capacity of steel columns, not wood, and thus A is the factor not used in the determination of the capacity of wood columns.

156. A. In composite design, a concrete slab is connected to a steel beam with shear connectors that can develop the ultimate capacity of the concrete or the steel, whichever is less (D). Because the concrete and steel act together, a smaller size steel beam may be used than in conventional steel framing, which generally results in a more economical system (B). However, when a smaller steel beam is used, deflections tend to become greater, and thus more critical, in composite design than in conventional steel framing (C). Conventional steel framing can always be designed to carry the required loads, and thus A is the incorrect statement we are looking for.

157. C. Candidates are not generally expected to memorize complex formulas. If a deflection problem of this kind appears on the exam, beam diagrams from the AISC Manual will probably be reproduced in the Exam Information Manual. For a simple beam supporting a uniform load, the deflection formula is

$$\Delta = \frac{5w\,l4}{384EI}$$

Since wl = W, the total load on the beam, the formula becomes

$$\Delta = \frac{5W\,l3}{384EI} = 5(2000)(28)(28 \times 12)^3$$

$$= 0.52"$$

Note that the length of 28 feet must be converted to (28×12) inches before being cubed.

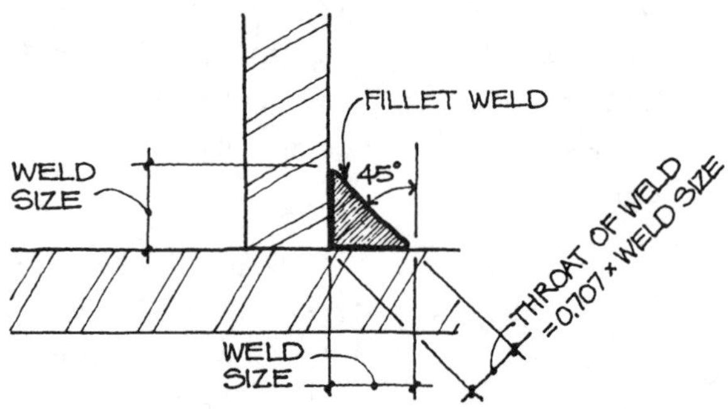

158. C. In fillet welds, the stress is always considered to be shear stress on the minimum throat area, regardless of the direction of the applied load, and the throat area is assumed to be equal to 0.707 times the weld size. The correct answer is C.

159. C. Reinforcing steel opposite the tension side of a reinforced concrete beam is called

compressive reinforcement, as it is located in the compression side of the beam and resists compression (I is correct, III is incorrect). The deflection of concrete members increases with time, beyond the initial deflection. This additional deflection, caused by shrinkage and creep, can be reduced by using compressive reinforcement (IV). As with tensile longitudinal reinforcing, the compressive reinforcement does not resist shear stress (II is incorrect). As only I and IV are true, the correct answer is C.

160. B. The formulas used to design columns are based on idealized pin-ended conditions, where the ends of the column are free to rotate, but not to move laterally (translate). Actual building columns, however, do not always meet these conditions: their ends may be free to rotate, or they may permit no rotation. Also, their ends may be free to translate, or they may be fixed against translation. To allow the column formulas to be used for all end conditions, the factor K was devised. This factor is multiplied by the actual unbraced length 1 to arrive at the effective length Kl, which is then used to design the column. B is therefore the correct answer.

161. B. A moment-resisting frame, as the name suggests, resists forces by developing moments at the joints between beams and columns. The moments in the beams produce tension in one beam flange and compression in the opposite flange. To transfer these forces from the beam flanges to the columns, the flanges must be rigidly attached to the columns by plates welded to the columns and bolted to the beams, as shown in I, or the beam flanges must be directly welded to the columns, as shown in III. The shear forces in the steel beam webs are transferred to the columns by plates welded or bolted to the beam webs and welded to the columns, as shown in I and III. Steel beams that are not part of moment-resisting frames are simply supported, and need only have the shear in their webs transferred to the column, either by a beam seat, as shown in II, or a plate, as shown in IV. The top angle in II serves only to hold the top of the beam in place, but is too flexible to rigidly attach the beam to the column.

162. D. The midspan deflection is equal to the sum of the deflection caused by the uniform load and the deflection caused by the concentrated load. The deflection formulas may be found in a number of reference sources, including the AISC Manual.

The uniform load deflection =

$$\frac{5}{384}\frac{w14}{EI} = \frac{5}{384} \times \frac{200(30)(30 \times 12)^3}{29(106)(2,100)} =$$

0.599"

The concentrated load deflection =

$$\frac{P1^3}{48EI} = \frac{9,000(30 \times 12)3}{48\,(29)\,(10)6\,(2,100)} = 0.144"$$

The total deflection = 0.599" + 0.144" = 0.743" (answer D).

Notice that the length of 30 feet must be converted to (30 × 12) inches before being cubed.

163. D. There are a number of ways to reduce the deflection of a reinforced concrete beam. The moment of inertia I of the beam may be increased by making the beam wider or deeper, the modulus of elasticity E of the concrete may be increased by using higher strength concrete, or additional reinforcing steel may be used. Of these, the most efficient way to reduce the deflection is to deepen the beam (choice D), because the I value of the beam is proportional to the third power of the depth. For example, doubling the depth increases the value of I eightfold ($2 \times 2 \times 2$), which would reduce the deflection to 1/8 of its original value.

164. C. In this question, referring to the AISC Manual is helpful, but not necessary, to solve this problem. In the Manual, the maximum moment for a beam with a uniformly increasing load is given as .1283 Wl. In this case, W = 1,200 × 24/2 = 14,400#, and M = .1283 (14,400)(24) = 44,340'# (correct answer C). Without the Manual, one would determine the left reaction by taking moments about the right end.

$$-1,200(24/2)(24/3) + R_L(24) = 0$$

$$R_L = 4,800\#$$

The maximum moment occurs where the shear is equal to zero. Since the load increases at the rate of 1,200/24 = 50#/', we determine the distance x from the left end to the point of zero shear, as shown below.

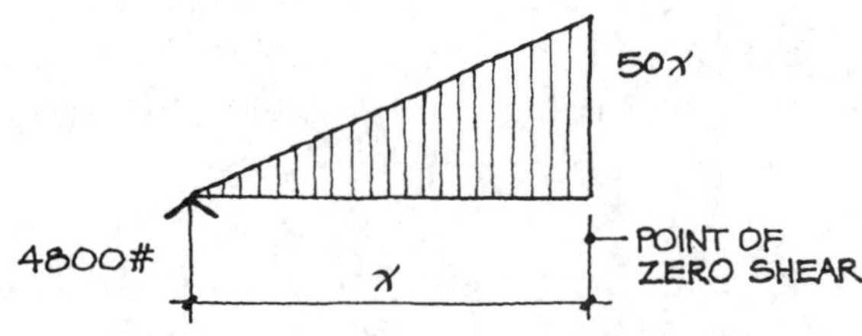

$$50\,(X)(x/2) = 4,800$$

$$x^2 = 4,800/25 = 192$$

$$x = \sqrt{192} = 13.856'$$

The maximum moment = 4,800(13.856) − 4,800(13,856/3) = 44,340'#, the same as we determined previously.

165. B. Reinforcing bars are furnished with rolled-in markings. These identify the producing mill (Y in this case), the bar size (number 4), the type of steel (N for new billet), and an additional marking for higher-strength steels (60 for 60 ksi yield strength). The bar size represents the diameter of the bar in eighths of an inch, thus 4/8 is 1/2 inch.

166. A. See the explanation for answer 169.

167. C. See the explanation for answer 169.

168. D. See the explanation for answer 169.

169. B. Questions about the stress-strain diagram have appeared on past exams, and candidates are therefore urged to become familiar with it. If we test a steel bar in tension in a testing machine and plot unit stress vs. unit strain, we will arrive at a diagram like the one shown on page 34. Hooke's Law states that up to the elastic limit (point A), unit stress is directly proportional to unit strain—in other words, the stress-strain diagram is a straight line. The constant ratio of unit stress to unit strain in this region, which can also be expressed as the tangent of angle θ, is called the modulus of elasticity. As we continue to test the bar, we reach a point where the bar continues to stretch with no increase in load, and the unit stress at which this occurs is called the yield point (point B). If we continue to increase the load on the steel bar, we will eventually reach the maximum unit stress that can be developed before it fractures. This stress is called the ultimate strength (point C).

170. B. From the table on page 64, the maximum spacing of lines of bridging for a 36LH07 is 11'-0". The number of bridging spaces is equal to the joist span divided by the maximum spacing = 60/11 = 5.45.

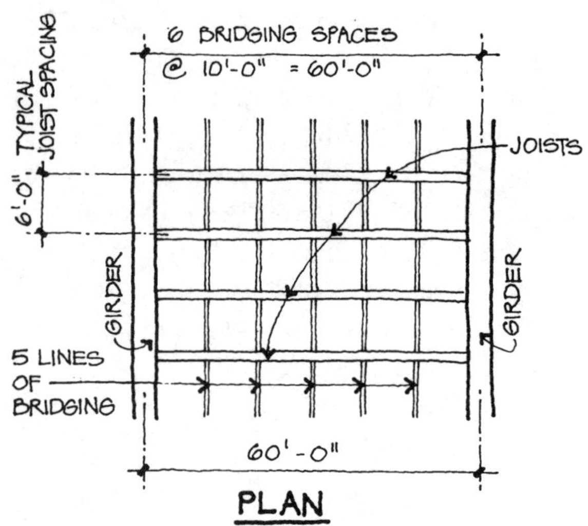

Because this must be an exact number, we use 6 spaces, or 5 lines of bridging, as shown in the sketch above.

171. A. I and III are true statements. II is also correct; truss chord members are normally subject to axial stress only, unless they support loads between the panel points, which causes bending. IV is incorrect, because the optimum depth to span ratio is about 1/10. V is also incorrect; increasing the depth of a truss decreases the stress in the chord members, but the stress in the web members depends on their slope, not on the truss depth.

172. B. For spans longer than 150 feet, steel arches are often trussed, to increase their bending resistance (A is false, B is true). C is also incorrect: the horizontal thrust is unaffected by the trussing, and depends only on the load, span, and rise. D is irrelevant; all types of arches may be fixed, two-hinged, or three-hinged.

173. B. Solving this kind of problem is very easy if you know how to use the tables. The total load supported by each joist is equal to $(20 + 20)6 = 240$ pounds per linear foot. The live load per joist is equal to $20 \times 6 = 120$ pounds per linear foot. We enter the table with a span of 60 feet, and read the figures "266" and "153" corresponding to the joist designation 36LH07. This means that the joist can safely support a total load of 266 pounds per linear foot, and that a live load of 153 pounds per linear foot will produce a deflection of 1/360 of the span. Since the actual total load is 240 pounds per linear foot, which is less than 266, and the actual live load is 120 pounds per linear foot, which is less than 153, the 36LH07 joist is satisfactory (correct answer B). While the 36LH08 and 09 can also support the load, they are both heavier than the 36LH07. The 32LH06 may not be used, since it can only support 234 pounds per foot (less than the actual load of 240). Also, a live load of 119 pounds per foot will produce a deflection of 1/360 of the span (less than the actual live load of 120).

174. D. Trusses may be solved by a number of different methods. In the method of sections, which works very well in this case, the truss is cut by an imaginary section. A free body diagram is drawn of the portion of the truss isolated by cutting the section. Moments are taken about the intersection of two of the cut members to determine the force in the third member.

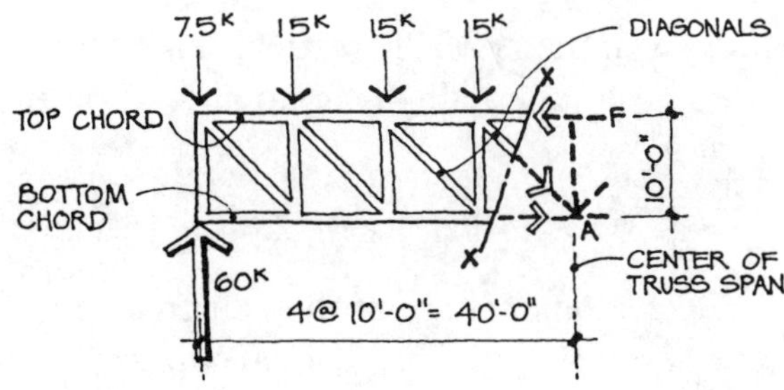

For a parallel chord truss, as in this problem, the internal force in the chord members is maximum at the center of the truss span. We therefore cut an imaginary sec-

tion x-x as shown in the illustration and take moments about point A, where the bottom chord and diagonal intersect.

$\Sigma M = 0$
$60(40) - 7.5(40) - 15(30) - 15(20) - 15(10) - F(10) = 0$

$F = (2,400 - 300 - 450 - 300 - 150)/10$
$\quad = 1,200/10 = 120$ kips (correct answer D)

175. B. Concrete domes are stressed principally in compression. As they are generally very thin, they are not able to resist much bending moment. A and C are therefore incorrect. The forming of a concrete dome is expensive, and therefore D is also incorrect. B is the only correct answer.

176. C. A joist girder is a steel truss used to support equally spaced steel joists, which deliver concentrated loads to the top chord panel points. The load at each panel point is equal to the unit load (40 psf) multiplied by the joist spacing (6 feet) multiplied by the joist span (60 feet). Panel point load = $40 \times 6 \times 60 = 14,400\# = 14.4$ kips (C).

177. B. The closely spaced columns of layouts A and B considerably reduce the bending moment in both directions. The cantilevers on all four sides in layout B also reduce the moments at the center of the spans, making it the most economical arrangement.

178. D. The slenderness ratio of a steel column is the effective length of the column KL divided by its least radius of gyration r. The effective length KL is the actual unbraced length L multiplied by the factor K, which depends on the support conditions at the ends of the column.

179. D. When a flat roof deflects under load, a concave surface results. Rainwater can collect in such areas, increasing the deflection, which further increases the amount of ponded water, and so on. During intense rainstorms, therefore, ponding of flat roofs can result in problems, including even collapse. Because of their long span and relative flexibility, long-span joists are particularly vulnerable to this kind of trouble (A is incorrect). Therefore, the joists should be cambered or pitched sufficiently so that rainwater cannot build up (D is correct). Pitching the top chords of joists is one way to achieve slope, but is usually more expensive than using parallel chord joists that are sloped, or building in camber (B is incorrect). C is also incorrect: designing joists for a uniform amount of water is not required if the roof has sufficient slope, and may not be adequate if the roof is flat.

180. C. Choice I, conventional cast-in-place reinforced concrete, is not practical for a 62-foot span. Choice II, glued laminated beams, is not usually used for parking structures because of the required fire rating. Likewise, choice V would be unlikely because exposed steel girders have no fire rating. That leaves III and IV as possible choices, both of which are likely to be practical and economical. C is therefore the correct answer.

181. C. While there are exceptions, when one dimension of a typical bay is much greater than the other, the system generally requiring the least amount of material in the floor framing has short beams and long girders (scheme 2). However, the disadvantage of this system is the greater overall depth of construction required. Conversely, long beams and short girders (scheme 1) usually require more material, but result in less depth of construction. Answer C is correct.

182. D. Because long-span structures are more vulnerable to failure than conventional structures, their designs are sometimes checked by independent third party architects or engineers—so-called "peer review." This additional design check does not eliminate the need for checking by the design team (A) or the building department (C), nor does it relieve the design team of responsibility (B). It does, however, provide an additional safeguard against design deficiencies (correct answer D).

183. D. Statements I, II, and IV are correct. Statement III is incorrect. The largest part of the structural framing cost in a suspension roof is in the fittings, connections, and anchorage members. Statement V is also incorrect. The draped cable will assume the shape of a parabola. D is therefore the correct answer.

184. C. To calculate the horizontal thrust, we isolate the left half of the arch.

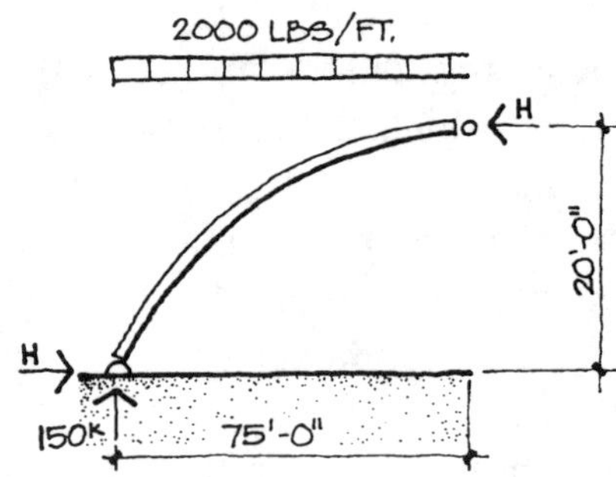

By symmetry, each vertical reaction is equal to one half of the total load on the arch

$$= 2{,}000 \times 150/2 = 150{,}000\# = 150 \text{ kips}$$

Take moments about the center hinge, where the moment is equal to zero

$$\Sigma M = 0$$

$$150(75) - 2.0(75)(75/2) - H(20) = 0$$

$$H = [150(75) - 2.0(75)(75/2)]/20$$

$$= (11{,}250 - 5{,}625)/20$$

= 5,625/20 = 281.25 kips (correct answer C)

185. D. All of the statements are correct except I. A composite beam consists of a steel beam and a concrete slab that are connected so that they act together as a single structural unit to resist bending stresses. A composite beam is much stiffer than a non-composite beam having the same depth, size, load, and span (III). However, because composite design usually allows us to use shallower beams (II), deflections should be checked (IV). The horizontal shear stress between the steel beam and the concrete slab, not moment, is resisted by shear connectors, such as studs, welded to the top flange of the beam and embedded in the concrete (I is incorrect).

186. A. The terms *ties* and *spirals* refer to the type of reinforcing used to wrap around the vertical reinforcing bars in reinforced concrete columns. Spirals are more effective than ties, and therefore spiral columns have greater axial load capacity than tied columns (I is correct). In plan, ties may be square, rectangular, or round, while spirals are always round. However, in both cases, the column may have any shape (III is incorrect). The use of ties or spirals has no effect on a column's fire resistance or forming (II and IV are incorrect).

187. C. Although most of the questions on the structural tests are conceptual, there are always some that require calculations, as in this case. Here we have a column that supports a roof and one floor. We first determine the column load per square foot, which is the total of the dead and live loads = (20 + 20 + 75 + 100) = 215 pounds per square foot. The total column load is equal to the load per square foot multiplied by the area tributary to the column = 215 ×

500 = 107,500#. The required footing area is equal to the total column load divided by the soil bearing value = 107,500 ÷ 3,000 = 35.83 square feet. We therefore use a 6'-0" × 6'-0" pad (choice C), which has an area of 36.0 square feet.

188. C. Long span structures generally have little redundancy, or alternative ways of resisting load (A is incorrect). In a long span structure, each connection supports a large area, so that the connections are more critical than in conventional structures (B is incorrect). C is correct: for a number of reasons, long span structures are more vulnerable to overall collapse, in case of accidental overload or a defect or weakness in a member or joint, than conventional structures. D is incorrect; although certain types of long span structures have a measure of instability, there is no requirement for a dynamic load analysis.

189. B. A beam with compression reinforcement, in addition to tension reinforcement, is known as a doubly-reinforced beam. Compression steel may be used in cases where the concrete alone is inadequate to resist the compressive stress, or where it is desirable to reduce long-term deflection due to creep and shrinkage.

190. C. First of all, what is a circular dish roof? As implied by its name, it is a circular roof structure curving upward in all directions from its lowest point at the center (V is correct, VI is incorrect). A number of radial cables, which are in tension, are stretched between the outer ring and the inner ring. The cables pull outward on the inner ring, putting it in tension (IV). At the same time, they pull inward on the outer ring, compressing it (I). See sketch on the following page.

191. C. Lateral bracing of trusses must be provided to prevent the compression members from buckling (I). Truss members are stressed principally in tension or compression only (IV is incorrect). However, small secondary bending and shear stresses are produced by the restraint against rotation at the joints. These secondary stresses are often ignored for conventional trusses, but should be considered for long span trusses (II). Trusses are always made up of triangles because the triangle is inherently a very stable shape; it is the only geometric figure that cannot be changed in shape without changing the length of one or more of its sides (III).

192. B. To solve this simple beam problem, you must know (1) the maximum moment in a beam supporting a concentrated load at midspan is PL/4, (2) the required section modulus S is equal to M/F_b, and (3) the allowable bending stress F_b for ASTM A36 steel is $0.66\,F_y = 0.66(36) = 24.0$ ksi. Thus, $M = PL/4 = 10(30)/4 = 75.0'$ k. $S = M/F_b = 75 \times 12/24 = 37.5$ in.^{3}. Note that we multiply the moment in ft.-kips by 12 to convert it to inch-kips, so that the section modulus comes out in inches^3.

193. D. A space frame consists of a series of parallel and perpendicular trusses that are connected at their points of intersection to form a two-way system (IV). Since all the members of the space frame resist the applied loads (III), the space frame is stiffer than the one-way trusses (I) and consequently can be made shallower than the one-way trusses (II).

194. D. Deflection is the movement that a beam undergoes when subject to load. The immediate deflection is determined by four factors: the load (I), the span (III), the stiffness of the beam material (measured by its modulus of elasticity), and the moment of inertia of the beam (V). The modulus of elasticity of concrete is a function of its weight and its strength (VI). Concrete members deflect when the load is initially applied and then continue to deflect with time (II). The yield point of the reinforcing steel (IV) has no effect on the moment of inertia, and therefore the deflection, of the beam section.

195. A. Concrete has a certain amount of shear strength. However, if that shear strength is insufficient to resist the shear stress, web reinforcement in the form of stirrups must be added. Of the other choices, B describes compressive steel, C generally refers to top bars over supports, and D describes ties used to laterally brace compressive steel.

196. B. Short of looking it up in the beam diagram section of the AISC Manual, how does one answer this question? Intuitively, one should know that because the loading increases towards the right end of the beam, the right reaction must be greater than the left reaction. Consequently, the shear at the right end is greater than at the left end. C incorrectly shows the shear at the left end to be greater than at the right, while D incorrectly shows zero shear at the left end. That leaves A and B as possible choices. If the load were uniform, the shear would change at a uniform rate and so the shear diagram would be a straight line. But because the load increases towards the right end, the shear changes at an increasing rate, as shown correctly in B. A shows the shear changing at a decreasing rate, which is incorrect.

197. D. A rigid frame can be made of structural steel with welded joints (B), but it can also have bolted joints, or be made of reinforced concrete. It can have fixed bases (C), but it can also have hinged bases. The best, most inclusive definition is that in choice D. A is not a rigid frame at all, but rather a braced frame.

198. B. In a reinforced concrete beam, if crushing of the concrete and yielding of the steel occur simultaneously, it is called a balanced design. In order to assure that the steel yields before the concrete crushes (III is correct), the amount of reinforcement may not exceed 75 percent of that which would produce a balanced design (I is incorrect, II is correct).

199. B. Steel columns bear on and are usually welded to steel base plates, which spread the column load over the supporting foundation. The bearing pressure is equal to the column load divided by the area of the base plate = 200,000#/16" × 16" = 781 psi.

200. D. All three methods of splicing reinforcing bars are allowed. Lapped splices are usually the least costly and therefore the most common. Welded splices are also used extensively. There are several types of mechanical connection devices available for splicing reinforcing bars. In general, these devices are proprietary and not as widely used as the other two methods.

201. C. A shell, or thin shell, is a structure with a curved surface that resists load by compression, shear, or tension in its own plane (I is incorrect), but which is too thin to resist any appreciable bending stresses (III is correct). Shells are strong in resisting uniform loads, but cannot resist any substantial concentrated loads, which tend to induce bending (II is incorrect).

202. A. The reference screens furnished to candidates at the exam included the beam formulas for this condition, one would immediately see that the reaction at A is 7wL/16 = 7(1.0) (24)/16 = 10.5 kips. But can one solve this problem if the formulas are not given? Yes, pretty easily. We isolate span AB, as shown below.

The moment at B is equal to the reaction at C times the length of span BC = 1.5 kips × 24 feet = 36.0 ft.-kips, as shown. Take moments about B.

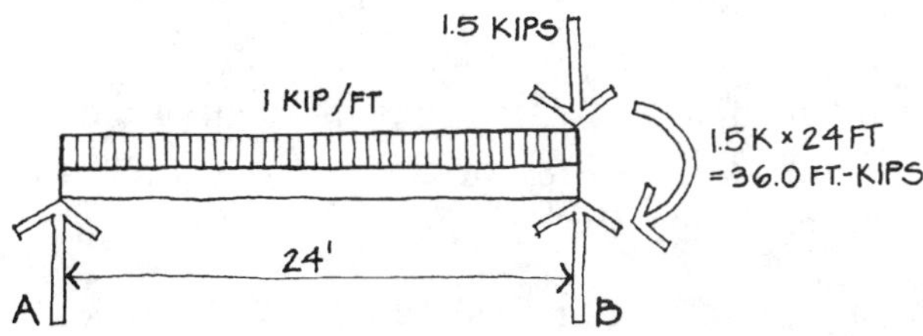

$$\Sigma M_B = 0$$

$$36.0 + R_A(24) - 1.0(24)(24/2) = 0$$

$$R_A(24) = 24(12) - 36$$

$$R_A = [(24)(12) - 36]/24 = 12.0 - 1.5$$

$$= 10.5 \text{ kips}$$

203. C. Composite beams, steel plate girders, and wood I-joists are examples of built-up members in which the flange-to-web connections must resist horizontal shear stress.

204. A. It is unnecessary to check the strength of wood beams in shear perpendicular to the grain, because the member will always fail first in shear parallel to the grain. A is therefore the incorrect statement we are looking for. B is correct: in all beams, the horizontal shear stress is always equal to the vertical shear stress. C is also correct: the allowable tension or compression parallel to the grain is many times greater than the allowable horizontal shear stress (parallel to the grain). D is a correct statement; for example, if a 12" deep beam is notched 2" on its bottom face at the end, the actual shear stress is increased about 50 percent. In detailing wood construction, therefore, notches should be avoided.

205. A. The modulus of elasticity of concrete is approximately w33 $\sqrt{f_c'}$, where w is the unit weight of the concrete in pounds per cubic foot and f_c' is the compressive strength of the concrete in psi. Thus, as the unit weight (I) or the strength (II) increases, the modulus of elasticity increases. The amount of reinforcing steel (III) is irrelevant, and the depth of the member (IV) affects the moment of inertia of the member, not the modulus of elasticity of the material.

206. D. Reinforcing steel must be protected against corrosion and fire by an adequate amount of concrete cover. The more severe the exposure (for example, concrete cast against earth), the greater the required concrete cover.

207. B. The margin of safety for reinforced concrete is provided by the load factor and ϕ, the strength reduction factor. The load factor, which is 1.4 for dead load and 1.7 for live load (A is correct), is based on the possibility that sometime during the life of the structure, the service loads may be exceeded (C is correct). The strength reduction factor ϕ allows for variations in material strengths and actual construction dimensions, as well as inaccuracies or approximations in design calculations (D is correct). Depending on the type of stress,

ϕ varies from 0.70 to 0.90 (B is the incorrect statement).

208. C. The base plate tends to bend as shown and must be thick enough so that the allowable bending stress is not exceeded.

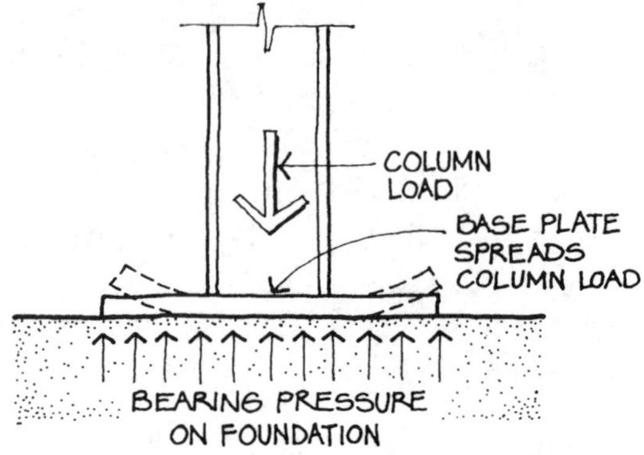

209. B. Open web steel joists are standardized lightweight steel trusses that are shop-fabricated (A and D are incorrect). C is also incorrect: longspan joists can be used for spans up to 96 feet, and deep longspan up to 144 feet. B is correct; there are numerous steel joist assemblies that have fire-resistance ratings, but they all require additional materials to provide fire protection.

210. C. Truss I is a Vierendeel truss, which is actually a type of rigid frame, while truss II is a typical triangulated truss. For the same load, span, and depth, the axial stresses in the chords of both trusses are equal (A is incorrect). The absence of diagonals in truss I does not cause instability (B is incorrect), but it does require the chords to resist shear, which results in bending moments in the chords. The verticals are also subject to bending moments (C is correct). Vierendeel trusses also tend to have high deflection (D is incorrect).

211. A. A Vierendeel truss has no diagonals. Therefore, the web is unencumbered (I) and the joints have a simple appearance

(II). However, Vierendeel trusses tend to have high deflection (III is incorrect). Also, as the members must be adequate to resist both bending moment and direct stress, a Vierendeel truss requires much more material than a triangulated truss (IV is incorrect).

212. A. The diagram shows a system in which glulam 2 is suspended between the cantilevered ends of glulams 1 and 3. Connections A and B therefore transfer the vertical reactions at the ends of glulam 2 to glulams 1 and 3. This system has smaller maximum moments than a comparable simple span system, thereby permitting longer spans and/or larger loads for a given size member. The suspended member (glulam 2) can also be made shallower because of its reduced span.

213. A. This is a problem involving number crunching and conversion factors. In the deflection formula, because $wL = W$ (# per ft. × ft. = total #), the formula becomes $\Delta = 5WL^3/384EI$, where $\Delta = L/360 = 24$ ft. × 12 in./ft./360 = 0.8"

$W = 1{,}000\#/\text{ft.} \times 24 \text{ ft.} = 24{,}000\#$

$L^3 = (24 \text{ feet} \times 12 \text{ inches/foot})^3$

$E = 29{,}000{,}000 \text{ psi}$

$0.8 = 5(24{,}000)(24 \times 12)^3/384 \times 29{,}000{,}000 \times I$

$I = 5(24{,}000)(24 \times 12)^3/384 \times 29{,}000{,}000 \times 0.8$

$= 321.8 \text{ in.}^4$ (answer A).

214. D. A joist girder is an open web steel truss used as a primary framing member. It is designed as a simple span member supporting equally spaced concentrated loads from a floor or roof system, generally consisting of open web steel joists. These concentrated loads are considered to act at the panel points of the joist girder. Joist girders are standardized for various depths and spans.

215. C. This question tests candidates' conceptual understanding of trusses. A truss is analogous to a steel beam; for a given load and span, the deeper the beam, the smaller the tensile and compressive forces in the flanges. Likewise for a truss: the deeper truss B has smaller chord forces than the shallower truss A (I is correct). From $\Sigma V = 0$, the vertical component of the force in the first diagonal of each truss is equal to R. For truss A, the horizontal component is also equal to R, as the truss depth and the panel length are both 10 feet. The total force in the diagonal is $\sqrt{R^2 + R^2} = 1.41R$. For truss B, the horizontal component is $10R/15 = 0.67R$, as the truss depth is 15 feet and the panel length is 10 feet. The total force in the diagonal is $\sqrt{R^2 + (0.67R)^2} = 1.20R$. Therefore, the diagonals in truss B have lower axial tension forces than those in truss A (II is incorrect). Finally, we cut a section through the first vertical of each truss, as shown on the following page. From $\Sigma V = 0$, the force in the vertical of each truss is equal to R (III is correct).

216. C. The flat plate system (A) has inexpensive formwork, but it is not generally economical for heavy loads or spans longer than 25 feet. The flat slab system (B) is economical for heavy loads, but not when the spans are much longer than 25 feet. The pan joist system (D) might be economical in this case, but it is generally more appropriate for moderate loads and spans where the bays are rectangular. The waffle slab system (C) is appropriate for heavy loads, square bays, and spans up to about 40 feet.

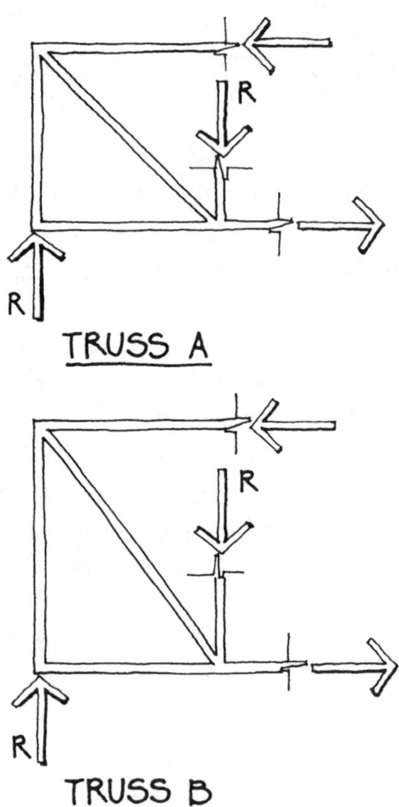

217. C. The combined effect of prestress and the applied loads usually results in compression over the entire cross section of a beam, which prevents tension cracks (A) and allows smaller sections to be used (D). Offsetting these advantages are greater labor and material costs (B). C is the incorrect statement we are looking for: pretensioned members are usually produced at a casting yard away from the building site, while posttensioned members are generally cast in place.

218. D. From the beam diagrams on page 69, we read that the maximum deflection Δ of a cantilever beam with a concentrated load at the free end is equal to $PL^3/3EI$. Substituting, we have $0.67 = 10,000(10 \times 12)^3/3 \times 29 \times 10^6 \times I$. $I = 10,000(10 \times 12)^3/3 \times 29 \times 10^6 \times 0.67 = 296.4$ in[4]. Notice the units: 10 kips becomes 10,000 pounds, and

10 feet becomes (10×12) inches. In that way, I will be expressed in inches[4].

219. D. In the stub girder system, steel floor beams sit on top of a stub girder, instead of framing into a girder. Because the beams are clear of the girder, they may be designed for simulated continuity (IV). The space between floor beams can be used for the mechanical and electrical distribution systems. Short lengths of filler beams the same depth as the floor beams are welded to the top of the stub girder to provide a connection between the girder and the slab for composite action. The main advantages of the stub girder system are reduced weight of steel (I), reduced story height (II), and simplified steel erection (III).

220. A. A flat slab is a two-way concrete slab supported directly by columns, without the use of beams or girders. In this system, a portion of the slab at the columns is thickened, which is termed a drop panel. The tops of the columns are flared, and these are known as column capitals. Both drop panels and column capitals are used to reduce shear stress in the slab near the columns and to provide greater effective depth for negative bending moment.

221. C. In the staggered truss system, story-high trusses spanning the full width of the building are arranged in a staggered pattern: trusses in the odd-numbered column lines are erected in a given story, while those in the even-numbered column lines are in the stories above and below. Thus, the floor system is supported alternately by the top chord of one truss and the bottom chord of the next. This provides an efficient framing system for the tall, narrow buildings typically used for hotel and residential occupancies (III is correct). The staggered truss system is not generally economical for buildings of less than 8 or 10 stories (II is incorrect). Similarly, the system is usually not efficient for spans less than about 45 feet (I is correct).

222. D. Because the allowable compressive stress depends on the slenderness ratio KL/r, the section with the greatest value of r will have the greatest allowable compressive stress. Closed sections, such as pipes and steel tubes, usually have greater r values than wide flange sections (B) or sections built up from angles (A and C). Note that in the slenderness ratio KL/r, we always use the least value of r if the value is different in each direction.

223. B. This question tests your ability to use bolt tables in the AISC Manual. The bolts have one value in shear and a different value in bearing, and the lower of these two values governs. Because there are two angles, the bolts are in double shear. From Table 1-D on page 62, each bolt has an allowable load in shear of 18.6 kips. From Table 1-E on page 63, each bolt has an allowable load in bearing on the 3/8" plate of at least 19.6 kips. The shear value governs, as it is lower. The number of bolts required is 50 kips/18.6 kips per bolt = 2.69; therefore we require 3 bolts.

224. A. This problem involves a two-step solution. First, the beam is considered to expand freely due to the temperature change, and the amount of this expansion is calculated. Second, because the beam cannot expand, we calculate how much compressive stress is required to push the beam back to its original length. Expansion due to temperature change = coefficient of expansion × length × temperature change = $(6.5 \times 10^{-6}) \times (24 \times 12) \times (100 - 40) =$ 0.11232". (Note that we multiply the length of 24 feet by 12 to convert it to inches.)

How much compressive stress is needed to push the beam back to its original length, that is, to shorten it 0.11232"? Unit stress divided by unit strain = modulus of elasticity E, or $(P/A) \div (\Delta/L) = E$, or $P/A = E\Delta/L = (29 \times 10^6) \times (0.11232)/24 \times 12 =$ 11,310 psi compression.

225. B. The connection shown is a seated beam connection, in which the load is transferred from the beam to seat angle B by bearing, and then through bolts F to the column. Bolts E connect the beam to the seat angle, but do not resist any calculated load. Top angle A and bolts C and D serve to stabilize the beam, not support it.

226. A. Here we have a typical moment connection, as might be used in a rigid frame. The beam moment is basically resisted by the flanges, while the shear is resisted by the web. Therefore, the moment is transferred to the column by the flange plates, which are welded to the column and bolted to the beam flanges. The shear is transferred to the column by the web plate, which is welded to the column and bolted to the beam web.

227. B. Questions about structural costs sometimes appear on the exam. I is incorrect: moment connections are more expensive than non-moment or simple connections, and should therefore be used only when necessary. II is correct; rolled sections should be used whenever possible, as they are generally less expensive than built-up sections. For economy, a good rule is to use as few members as possible (III is correct). High-strength steel has a higher unit cost than conventional A36 steel, and therefore its use is not generally economical except for highly-stressed members (IV is incorrect).

228. B. The allowable shear in a fillet weld is based on the throat dimension, as shown.

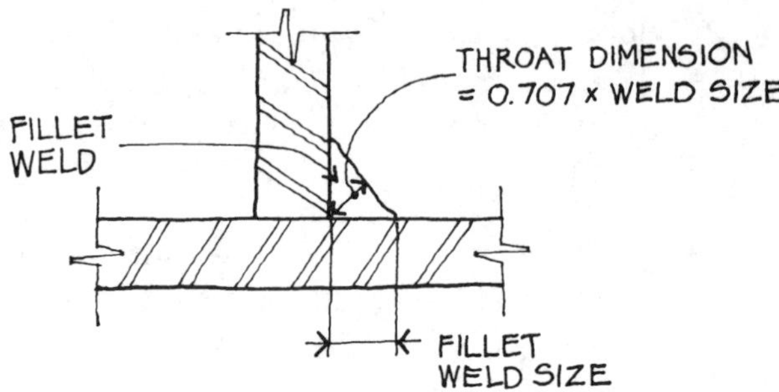

Thus, for a 1/4 inch fillet weld, the allowable shear per inch of weld = 0.25" × 0.707 × 21.0 ksi = 3.71 kips per inch. The total length of weld required is 80 kips ÷ 3.71 kips per inch = 21.6 inches. Half of this is required on each side of the beam web, or 21.6 ÷ 2 = 10.8 inches, which we round up to 11 inches.

229. B. Statement A is incorrect because reinforced concrete columns must be designed for a minimum amount of eccentricity, which is equivalent to bending moment, even if the load theoretically is axial. C is also incorrect; the strength reduction factor Ø is equal to 0.75 for spiral columns and 0.70 for tied columns, because of the greater toughness of spiral columns. Spiral columns are usually more expensive than tied columns because the spiral reinforcement costs more to fabricate (incorrect answer D). Only B is correct; a spiral column has about 14 percent more axial load capacity than a tied column with the same cross-sectional area and vertical reinforcement.

230. B. Although there have been few numerical problems on the exam concerning arches or other long span structures, candidates are expected to understand the concepts of such structures. An arch can resist both vertical and horizontal loads (D is incorrect). When an arch supports vertical loads, there is a horizontal reaction at each base, and the reactions are of equal magnitude and act in opposite directions (A is incorrect, B is correct). A pin or hinge in a structure cannot resist any moment, making C incorrect.

231. B. In a moment-resisting steel frame, the moments in the beams produce tension in one beam flange and compression in the other flange. To transfer these forces from the beam flanges to the columns, moment-resisting connections are used, which must have adequate strength and not allow any slippage. For these reasons, the beam flanges are usually attached directly to the columns with full penetration groove welds (C), or with cover plates welded to the columns and welded (D) or bolted with high-strength bolts in slip-critical connections (A) to the beam flanges. ASTM A307 machine bolts are generally not used for moment-resisting connections because they rely on bearing to develop their strength, and in so doing allow some movement. B is therefore the correct answer.

232. B. Materials expand when heated and contract when cooled. The amount of expansion or contraction is equal to the product of three factors: the coefficient of thermal expansion of the material, the length of the member, and the temperature change. Thus Δ(expansion) = (0.0000065) × (24 ft. × 12 in./ft.) × (90 − 60) = 0.056" (answer B).

233. D. The principal purpose of a base plate under a steel column is to spread the column load over a relatively large area of the concrete foundation, so that the bearing pressure on the concrete is not excessive (correct answer D). The thickness of the base plate is determined by assuming the bearing pressure to be uniform and considering the portions of the plate outside of the column to cantilever from the column edges.

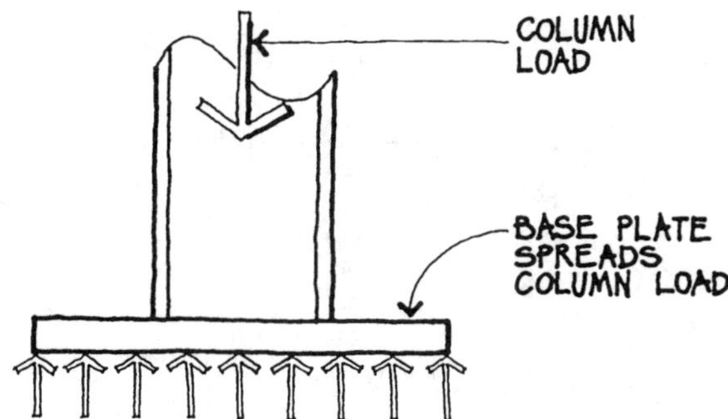

234. A. A Vierendeel truss has no diagonals. Therefore, large openings, such as doors and windows, may be made within the depth of a Vierendeel truss without conflicting with diagonal members (III is incorrect). A conventional triangulated truss tends to have less deflection and use less material than a comparable Vierendeel truss (I and II are correct). Both conventional and Vierendeel trusses can have loads applied between panel points (IV is incorrect), although this results in bending stress in the chord member. A is therefore the correct answer.

235. D. To transfer the flexural tensile and compressive forces in the beam flanges to the columns of a rigid frame, the flanges may be attached directly with full penetration groove welds (II) or with cover plates welded to the columns and bolted (III) or welded to the beam flanges. Two clip angles bolted to the beam web may be adequate to transfer shear, but are not adequate to transfer moment (I is incorrect). Likewise, a seat angle may transfer shear, but is inadequate to transfer moment (IV is incorrect). Therefore, the correct choices are II and III (answer D).

236. C. This problem requires a trial-and-error solution. Try a 6×8.

$l/d = (16 \text{ ft.} \times 12 \text{ in./ft.}) \div 5.5 \text{ in.}$

$= 34.9$

$(l/d)^2 = 34.9^2 = 1,218$

Allowable stress $= 0.30E/(l/d)^2$

$= (0.30 \times 1.6 \times 10^6) \div 1,218 = 394 \text{ psi}$

Required column area

$= 40,000\# \div 394 \text{ psi} = 101.5 \text{ sq.in.}$

The area of a $6 \times 8 = 5.5 \text{ in.} \times 7.5 \text{ in.}$

$= 41.25 \text{ sq.in.}$ A 6×8 is therefore inadequate.

Next we try an 8×8.

$l/d = (16 \times 12) \div 7.5 = 25.6$

$(l/d)^2 = 25.62 = 655.36$

Allowable stress

$= (0.30 \times 1.6 \times 10^6) \div 655.36$

$= 732.4 \text{ psi}$

Required column area

$= 40,000\# \div 732.4 \text{ psi} = 54.6 \text{ sq.in.}$

The area of an $8 \times 8 = 7.5 \times 7.5 = 56.25$ sq.in. The smallest column that may be used is therefore an 8×8, answer C.

237. B. The staggered truss system (correct answer B) consists of story-high trusses spanning transversely between exterior columns and arranged in a staggered pattern. It is often an efficient and economical structural system for high-rise apartment buildings.

238. C. The ideal steel column to resist buckling is one whose radius of gyration r is the same in both directions, such as a pipe column or tubular section, rather than a wide flange section (C is correct). All the other statements are incorrect. The buckling tendency of a steel column depends on its end conditions, its length, and its radius of gyration, not on its yield point (A). The maximum allowable slenderness ratio Kl/r is 200, not 50 (B). And if the value of r is different in each direction, as with a wide flange section, the lower value is used to compute Kl/r (D).

239. B. The allowable shear value of a 1" diameter A325-N bolt is 16.5 kips in single shear and 33.0 kips in double shear (see Table 1-D of the AISC Manual, which is on page 62). The 5 bolts connected to the web of the W24 × 76 are in double shear (5 × 33.0 = 165 kips), and the 10 bolts connected to the web of the W27 × 102 are in single shear (10 × 16.5 = 165 kips). We must also check bearing stress; from page 63, the bearing value of a 1" diameter bolt is 69.6 kips for 1 inch of thickness. Because the web thickness of the W24 × 76 is 0.440", each bolt has a bearing value of 69.6 × 0.440 = 30.6 kips, which is less than the shear value of 33.0 kips and therefore governs. The capacity of the connection is 30.6 kips per bolt times 5 bolts = 153 kips (answer B).

240. D. Exam questions involving trusses may be conceptual, or they may require calculations, as in this question. We first determine the value of each reaction. By symmetry, each reaction is equal to one-half of the total load on the truss = (10 + 10 + 10 + 10 + 10) ÷ 2 = 25 kips. We next cut a section through the truss panel that contains member a, as shown, assuming member a to be stressed in tension (pulling away from the joint).

$$\Sigma V = 0$$

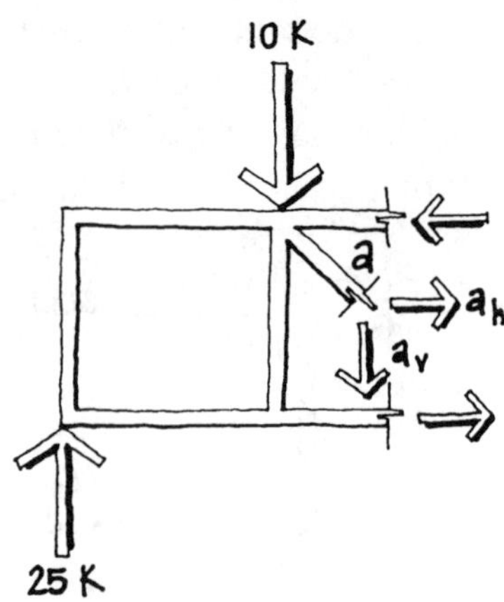

$$+ 25 \text{ kips} - 10 \text{ kips} - a_v = 0$$
$$a_v = +15 \text{ kips}$$

Since member a is inclined at 45°,

$$a_h = a_v = +15 \text{ kips.}$$

Therefore, the internal axial force in member $a = \sqrt{15^2 + 15^2} = 21.2$ kips. Since we assumed member a to be in tension, and the sign of its internal force comes out positive, our assumption is correct and member a is in tension (D is correct).

241. D. A space frame is a series of trusses that intersect in a grid pattern and are connected at their points of intersection. Space frames are often economical for enclosing large, square, column-free spaces, as in this question. The key to their economy is the use of repetitive members and connections (I). Other advantages include greater stiffness (II) and reduced depth (III). However, space frames are statically indeterminate and their structural analysis is complex (IV is incorrect). D is therefore the correct answer.

242. C. A drilled caisson bears on the soil at its bottom and is constructed by pouring concrete into a drilled shaft. The bottom of the shaft is often enlarged, or belled, in order to increase the bearing area and hence the bearing capacity of the caisson (correct answer C).

243. C. The shear capacity of a reinforced concrete beam depends on its width, its depth, and the ultimate 28-day compressive strength of the concrete (C is correct). If the shear capacity is insufficient to resist the shear force, web reinforcement may be added. The cross-sectional area of the longitudinal tension reinforcing is irrelevant in this regard (A is incorrect), and the beam's load and span affect the shear force on the beam, not its shear capacity (B and D are incorrect).

244. C. In this question, you must determine the maximum bending moment that can be resisted by a 6×12 wood beam. Using the formula $F_b = M/S$, we rearrange the terms so that $M = F_b S = 1,600\#/in.^2 \times 121.229$ in.^{3} = 193,966 in.-lbs. We convert this to ft.-lbs. by dividing by 12. $193,966 \div 12 =$ 16,164 ft.-lbs. (answer C).

245. D. The problem of selecting a steel joist has come up on some exams, and solving such a problem is very easy if you know how to use the Steel Joist Institute tables, such as that on page 66. In this case, the total load supported by each joist is equal to $(20 + 30)$ lbs./sq.ft. $\times 6$ feet = 300 pounds per linear foot. The live load per joist is equal to $30 \times 6 = 180$ pounds per linear foot. You will note that each joist in the table has two numbers corresponding to each span: the upper number represents the total load in pounds per linear foot that the joist can safely support, and the lower number represents the live load in pounds per linear foot which will produce a deflection of 1/360 of the span. In order to satisfy both load and deflection criteria in this problem, we must find a joist whose upper number is at least 300 and whose lower number is at least 180, and further, we

must select the lightest such joist. For the 40LH12, we read 330 and 157, and the joist is therefore adequate for stress ($330 > 300$), but inadequate for deflection ($157 < 180$). For the 40LH13, we read 390 and 185, and for the 44LH13, we read 404 and 212. For the 48LH12, we read 336 and 191 for a span of 81 feet, and these numbers would be slightly higher for an 80-foot span. Thus, all three of these joists satisfy both load and deflection criteria. From the second column in the table, the 40LH13 and 44LH13 each weighs 30 pounds per linear foot, while the 48LH12 weighs 25 pounds per linear foot. The 48LH12 (correct answer D) is therefore the lightest joist that satisfies the criteria in the problem. Although this procedure may seem involved, in actual practice it can be done rather quickly.

246. C. Exam questions on reinforced concrete design are more likely to be conceptual that numerical. This question tests your understanding of several reinforced concrete design concepts. Failure due to crushing of the concrete is sudden and without warning, while failure due to yielding of the tensile steel is more gradual and gives adequate warning of approaching collapse. In order to assure that failure due to yielding of the steel takes place before failure of the concrete, the code sets an upper limit on the reinforcement ratio of 0.75 of the ratio that would produce a balanced design. A is therefore a correct statement and C is an incorrect statement. B and D are also correct; the reinforcing steel is generally assumed to resist all the tensile stresses, and the ultimate load factors are greater for live load than for dead load because a specified live load is more apt to be exceeded than dead load, which is fixed. As C is the only incorrect statement, it is the answer to this question.

247. D. This is a typical steel beam problem that you should be able to solve very easily. The maximum moment in a simple beam supporting a uniform load is equal to $wL^2/8$ (a formula you should probably memorize). Thus, maximum moment M = 1,800#/ft. $\times$ (30 ft.)$^2 \div 8$ = 202,500 ft.-lbs. We convert this to in.-lbs. by multiplying by 12 = 202,500 $\times$ 12 = 2,430,000 in.-lbs. The required section modulus S = M/F$_b$, where F$_b$ is the allowable bending stress (24,000 psi for ASTM A36 members with full lateral support). Therefore S = 2,430,000 $\div$ 24,000 = 101.25 in.^{3}. Using the table below, we locate the first group of beams which have a value of S equal to or greater than 101.25. The beam in boldface type at the top of the group is the lightest beam that is adequate, in this case a W24 $\times$ 55 (correct answer D). While this is a pretty simple procedure, an even simpler procedure is to locate the group of beams which have a value of moment M$_R$ equal to or greater than 202.5 and select the beam in boldface type at the top of the group, again a W24 $\times$ 55.

248. A. The stub girder system is a steel beam-and-girder system in which the floor beams sit on top of the main girders, rather than framing into them. Short lengths of stub girders the same depth as the floor beams are welded to the tops of the main girders to provide a connection to the slab for composite action (A is correct). The advantages of the stub girder system are reduced weight of steel and reduced story height. See page 120.

249. A. Unlike structural steel, which has a constant value of modulus of elasticity, the modulus of elasticity of concrete varies with its strength and unit weight (I is correct). A reinforced concrete beam continues to deflect after it reaches its initial deflection (II is correct). III is incorrect because adding compressive reinforcement reduces the creep and hence the long-term deflection of a reinforced concrete beam.

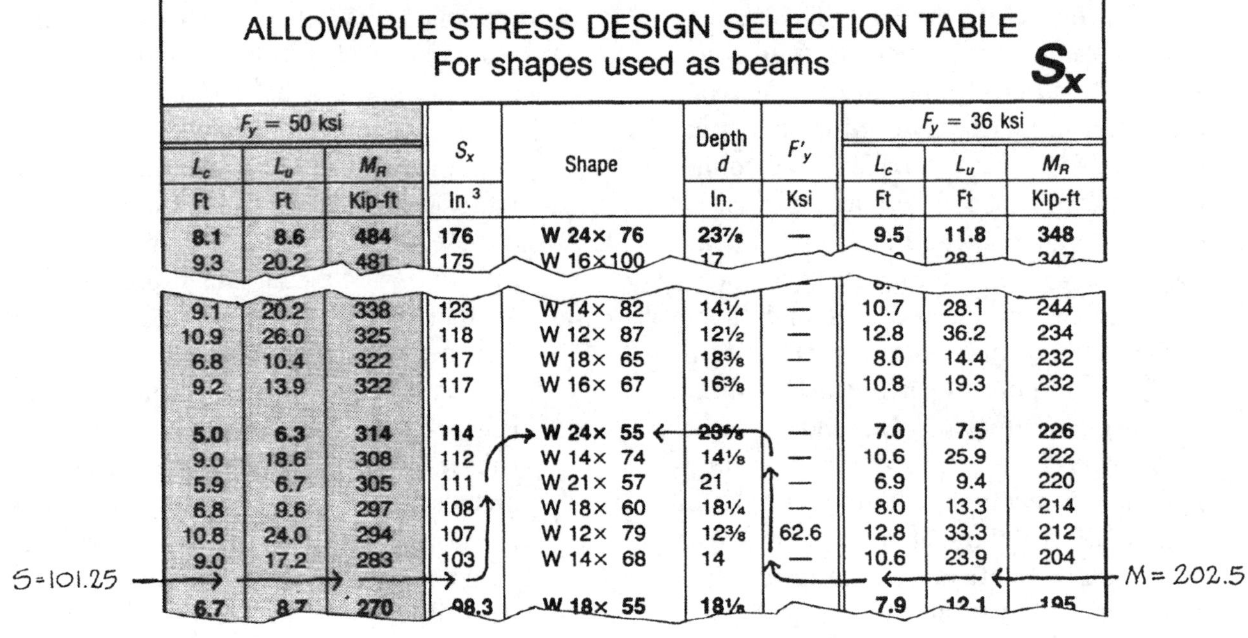

ALLOWABLE STRESS DESIGN SELECTION TABLE
For shapes used as beams — S_x

F_y = 50 ksi			S_x	Shape	Depth d	F'_y	F_y = 36 ksi		
L_c	L_u	M_R					L_c	L_u	M_R
Ft	Ft	Kip-ft	In.^{3}		In.	Ksi	Ft	Ft	Kip-ft
8.1	**8.6**	**484**	**176**	**W 24× 76**	**23⅞**	—	**9.5**	**11.8**	**348**
9.3	20.2	481	175	W 16×100	17			28.1	347
9.1	20.2	338	123	W 14× 82	14¼	—	10.7	28.1	244
10.9	26.0	325	118	W 12× 87	12½	—	12.8	36.2	234
6.8	10.4	322	117	W 18× 65	18⅜	—	8.0	14.4	232
9.2	13.9	322	117	W 16× 67	16⅜	—	10.8	19.3	232
5.0	**6.3**	**314**	**114**	**W 24× 55**	23⅝	—	**7.0**	**7.5**	**226**
9.0	18.6	308	112	W 14× 74	14⅛	—	10.6	25.9	222
5.9	6.7	305	111	W 21× 57	21	—	6.9	9.4	220
6.8	9.6	297	108	W 18× 60	18¼	—	8.0	13.3	214
10.8	24.0	294	107	W 12× 79	12⅜	62.6	12.8	33.3	212
9.0	17.2	283	103	W 14× 68	14	—	10.6	23.9	204
6.7	8.7	270	98.3	W 18× 55	18⅛		7.9	12.1	195

Since I and II are correct statements, A is the correct answer.

250. A. The truss in this question is called a *King Post truss*, in which the internal force in the vertical member is zero (correct answer A). To prove this, we isolate joint D and apply the basic equation of static equilibrium $\Sigma V = 0$.

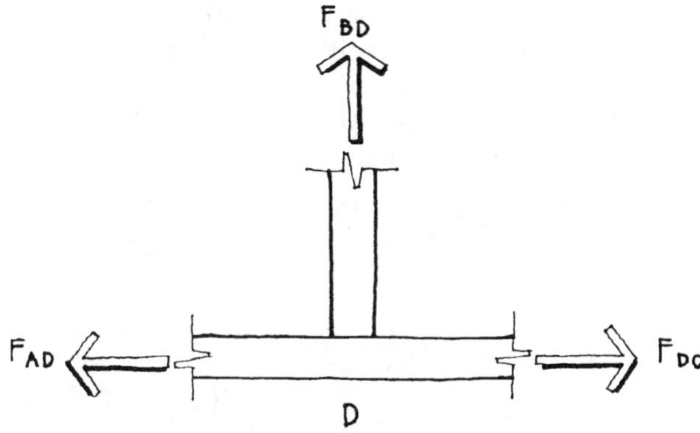

Since the only vertical force acting at the joint is $F_{B\text{-}D}$, it follows that $F_{B\text{-}D} = 0$.

251. C. In composite design, a steel beam and the concrete slab above it are connected so that they act together as a single structural unit to resist bending stresses. The concrete slab becomes part of the top flange and resists compressive bending stresses (C is correct). Connectors welded to the top flange of the steel beam resist shear, not flexural stresses (A is incorrect). Adding these connectors changes the beam from non-composite to composite and makes the beam stiffer. However, as composite design usually allows the designer to use shallower beams, it is necessary to check deflections (D is incorrect). Composite construction is most efficient with heavy loads and long spans (B is incorrect).

252. B. Stirrups are small U-shaped bars, such as #3 or #4, which are placed vertically in a reinforced concrete beam to reinforce the web where the shear stresses are excessive (correct answer B). A is incorrect, as compressive reinforcement consists of longitudinal bars placed in the compressive area of a beam to resist part of the compressive stress. Reinforcement is anchored by mechanical devices and/or embedment (C is incorrect), and lateral buckling of compressive reinforcement is prevented by the use of ties, not stirrups (D is incorrect).

253. D. The bolts in this question are in double shear, as there are two planes through each bolt that resist shear. From Table 23-I-F of the International Building Code, which is reproduced on page 67, the allowable load for each 7/8" bolt in double shear parallel to grain when the thickness of the main member is 3-1/2" (nominal 4") is 3,580#. The total load that can be transferred by two bolts is therefore $3{,}580 \times 2 = 7{,}160\#$ (correct answer D).

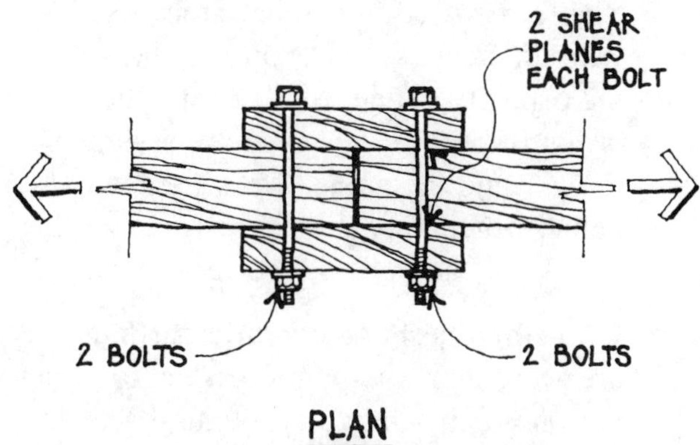

254. D. All steel, regardless of strength or other properties, has the same value of modulus of elasticity E, about 29,000,000 psi (D is correct). The other properties in this question—yield point, ultimate strength, and weldability—vary depending on the type of steel.

255. A. In this question, definitions of four different long-span roof systems are given, and you are asked to select the one which refers to a lamella roof. Answer A correctly defines a lamella roof. B is the definition of a space frame, and C and D describe systems that have been used but do not have specific names.

256. D. All four statements are correct (answer D). Air-supported membranes that enclose a space are pressurized by fans (I). Since the internal air pressure is slightly greater than the outside pressure, airlocks or special doors are required to get in and out of the space (II). Loss of pressure can cause the roof to deflate (IV); however, in that event, the velocity of the fans could be increased so that the deflation would take place over a long period of time. Large air-supported fabric structures are usually reinforced with steel cables (III).

257. C. The top chord of a simply supported truss is stressed in compression. As with all steel compression members, the capacity of the truss chord is based on its Kl/r value. K is usually assumed to be 1 and the value of r is the lower value, which may be with respect to either the *x-x* or the *y-y* axis (correct answer C).

258. C. In this question, four different diagrams are presented, and you are asked to select the one representing the flexural stresses in a reinforced concrete beam at failure. Diagram A represents the flexural stresses in a homogeneous rectangular beam, such as wood, and diagram B represents the flexural stresses in a reinforced concrete beam at working stresses. This is the basis of working stress design, which has been largely superseded by the strength design method. Diagram C correctly shows the flexural stresses in a reinforced concrete

beam at failure, which is the basis of strength design. The bunched arrows represent the compressive stress in the concrete and the single arrow represents the tensile stress in the reinforcing steel. Diagram D represents the flexural stresses in a steel beam when the entire beam profile is stressed to the yield strength. This is the basis of limit states design in steel, which is analogous to the strength design method used for reinforced concrete.

259. A. This truss problem can be solved in various ways, but the simplest way is to isolate the joint where the 20 kip load is applied and draw a free body diagram.

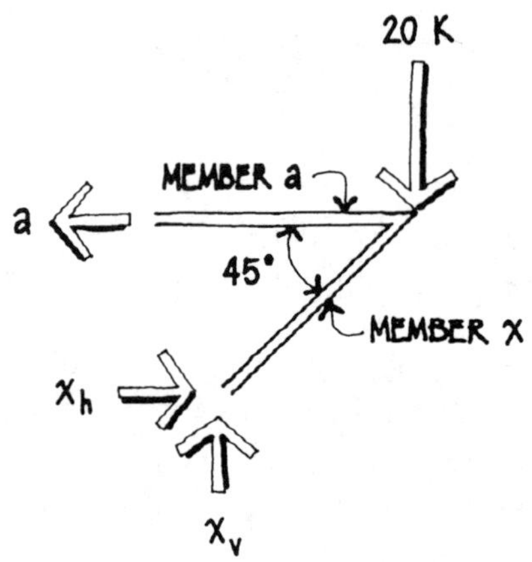

Assume member *a* is stressed in tension (pulling away from the joint) and diagonal member *x* is stressed in compression (pushing toward the joint).

$\Sigma V = 0$

$-20 + x_v = 0$

$x_v = +20$ kips

Since member x slopes at 45°, $x_h = x_v = 20$ kips

$\Sigma H = 0$

$+x_h - a = 0$

$a = +x_h = +20$ kips

Since all the signs come out positive, our assumptions about the directions of the stresses are correct, and member a is stressed in tension. A is therefore the correct answer.

260. B. A three-hinged arch is the only type of arch that is statically determinate, no matter what the loading is (B is correct). In the answer to Question 262, we show how the horizontal thrust at each end of a three-hinged arch is calculated.

261. C. When a flat roof deflects under load, a concave surface results. Rainwater can collect in such areas, increasing the deflection, which further increases the amount of ponded water, and so on. During intense rainstorms, therefore, the ponding of flat roofs can result in problems, including even collapse. Because of their long span and relative flexibility, long span open web joists are particularly vulnerable to this kind of trouble. Therefore, the joists should be cambered or pitched sufficiently so that rainwater cannot build up. Building pitch into the top chords of the joists is one way to achieve slope, but is usually more expensive than using parallel chord joists which are sloped, or building in camber (I is incorrect, II and IV are correct). Further a slope of 1/8 inch per foot is sometimes used, although this slope is considered minimum, it is considered acceptable practice (III is correct). C is therefore the correct answer.

262. C. A three-hinged arch is the only type of arch that is statically determinate. To calculate the horizontal thrust, we draw a free body diagram of the left half of the arch.

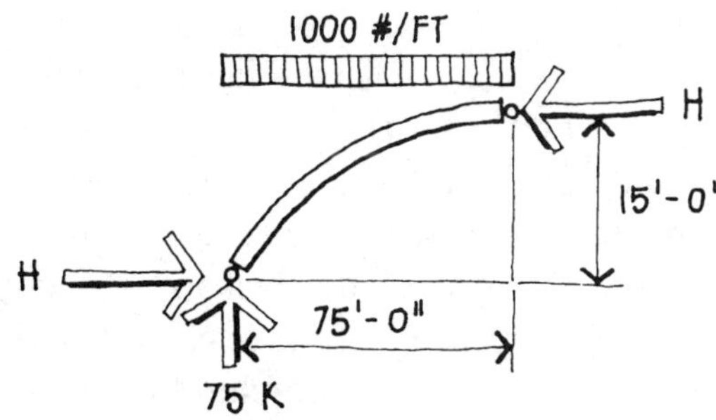

By symmetry, each vertical reaction is equal to one-half of the total vertical load on the arch = 1,000 lbs./ft. × 150 ft./2 = 75,000# = 75 kips. Take moments about the center hinge, where the moment is equal to zero.

$\Sigma M = 0$

$(75 \text{ kips} \times 75 \text{ ft.}) - (1.0 \text{ kips/ft.} \times 75 \text{ ft.}$
$\times \dfrac{75 \text{ft}}{2}) - (H \times 15 \text{ ft.}) = 0$

$H = [(75 \times 75) - (1 \times 75 \times 37.5)] \div 15$

$= (5,625 - 2,812.5) \div 15 = 187.5$ kips (answer C)

263. D. It is practically impossible to make butt joints sufficiently strong or permanent to adequately join laminations end to end. Instead, scarf joints (I and II); finger joints (III); or other similar joints must be used (D is correct).

264. B. The flat slab floor system is a two-way reinforced concrete system that is supported directly on the columns, generally without any beams or girders. A and D are therefore correct statements. The flat slab system is usually economical for heavy loads, as in warehouses (C is correct). B is incorrect and is therefore the answer to this question; flat slabs are relatively thin and are therefore not economical in reinforcing steel.

265. A. The exam invariably includes some questions that test candidates' familiarity with structural terms, as in this question. A composite deck is steel decking manufactured with deformations that mechanically bond the deck to the concrete slab above it, so that the deck and slab act together as a single structural element to span between floor beams (correct answer A). C and D are incorrect because they describe a composite beam, not a composite deck, and B is also incorrect.

266. D. Identification marks are rolled into the surface of reinforcing bars to denote the producing mill, the bar size, the type of steel, and the minimum yield strength. In this case, the producing mill has the symbol H, the bar size is #10, which is 1 − 1/4" in diameter, the type of steel is designated S, which means billet steel, and 60 means that the steel has a minimum yield strength of 60 ksi. The correct answer is therefore D.

267. D. I-shaped wood joists have a profile that looks like the illustration in the right column. As with any I-shaped member, such as a steel plate girder, the flanges substantially resist the flexural tension and compression, the web resists the shear, and the connections between the flanges and the web resist horizontal shear (D is correct).

268. A. Openings in beams have the least effect on the beam's load-carrying capacity if they are located in areas of low stress. The two main types of stress in beams are shear stress and bending, or flexural, stress. The shear stress is usually greatest near the supports and least near midspan. Thus, an opening in an area of low shear stress would be near the center of the span (A,

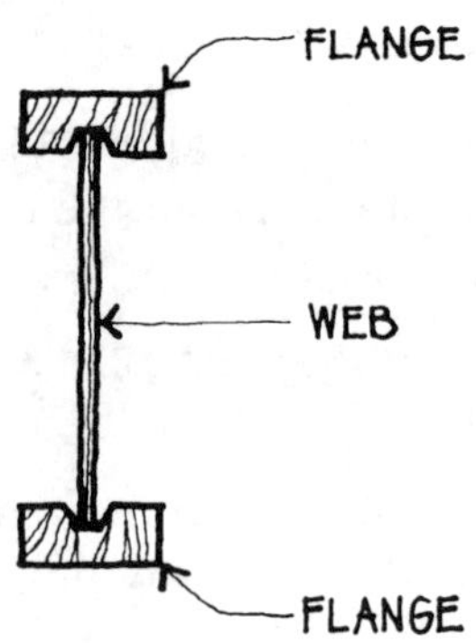

B, or C). For a simply supported beam, the bending stresses are greatest near the middle of the span and least near the supports. Within the beam depth, the bending stresses are greatest near the top (compression in the concrete) and bottom (tension in the reinforcing steel) and least at about the mid-depth of the beam. Therefore, an opening in an area of low bending stress would be near the supports (D) or at the mid-depth of the beam (A). The only location in an area of both low shear stress and low bending stress is A.

269. A. Prestressing of concrete is done by either pretensioning or posttensioning. In pretensioning, high strength steel is tensioned before the concrete is cast. After the concrete hardens, the prestress wires are cut and the prestress force is applied to the concrete through bond. Therefore, no end anchorages are required (B is correct). Most precast, prestressed members are pretensioned and again, no end anchorages are required. A is therefore the untrue statement we are looking for. In posttensioning, the steel tendons are stressed after the concrete is cast on the site, by jacking against anchorages at the ends of the member (D is correct). Whether pretensioned or posttensioned, prestressing results in more efficient use of the material, making smaller sections possible (C is correct).

270. A. The horizontal shear is transferred from the column to the footing by shear in the anchor bolts (correct answer D). The vertical load, not the horizontal shear, is transferred from the column to the footing by bearing of the base plate on the footing (A is incorrect). If there is movement in the column, it might be resisted by tension in the anchor bolts and bearing of the base plate on the footing (C is incorrect). Friction between the base plate and the footing is generally not considered to resist load (B is incorrect).

271. B. Shear failure of a beam is most likely to occur where the vertical shear is maximum, which is adjacent to the supports (II is correct, I is incorrect). Horizontal shear stress in a beam varies from zero at the outermost fibers to a maximum value at the mid-depth of the beam (III is correct, IV is incorrect). Because only II and III are correct, the answer is B.

272. A. On the structural test, you can expect to see a number of problems involving beams of various types (simple, cantilever overhanging) supporting various loadings. You may be asked to determine the beam reactions, shears, or moments. The best preparation for this kind of problem is to practice solving a number of beams, such as this one. First, what is the resultant of the triangular load?

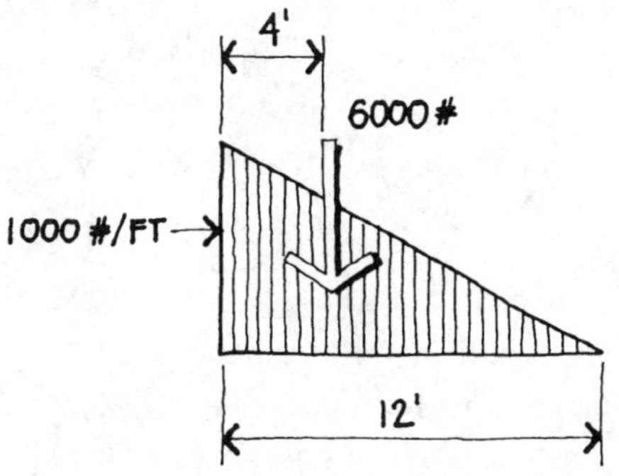

It is simply the area of the triangle, which is $(1,000\text{\#/ft.} \div 2) \times 12 \text{ ft.} = 6,000\text{\#}$, applied at the centroid of the triangle, which is 12 ft. $\div$ 3 = 4 ft. from the left end.

$\Sigma V = 0$

$+ V - 6,000\text{\#} - 5,000\text{\#} = 0$

$V = 11,000\text{\#}$

$\Sigma M = 0$

$(6,000\text{\#} \times 4 \text{ ft.}) + (5,000\text{\#} \times 12 \text{ ft.}) - M = 0$

$M = 24,000 + 60,000 = 84,000$ ft.-lbs.

Therefore, choice A is correct.

273. D. The load-carrying capacity of a wood column is determined by several factors: the modulus of elasticity E of the wood, which in turn depends on its species and grade; the allowable compressive stress F_c, which also depends on the species and grade of the wood; the ratio 1/d, where 1 is the unbraced height of the column and d is the least lateral dimension of the column; and the cross-sectional area of the column. Therefore A, B, and C are all correct. D is incorrect and therefore the answer to this question; the load-carrying capacity is not affected by the applied load. But the applied load may not exceed the load-carrying capacity.

274. C. A groove weld is placed between two butting plates or members and is usually stressed in direct compression or tension. A complete penetration groove weld is one whose depth is the same as the thickness of the member. The strength of a complete penetration groove weld is considered to be the same as that of the connected material (correct answer C).

275. D. The maximum size of coarse aggregate that may be used depends on the size of

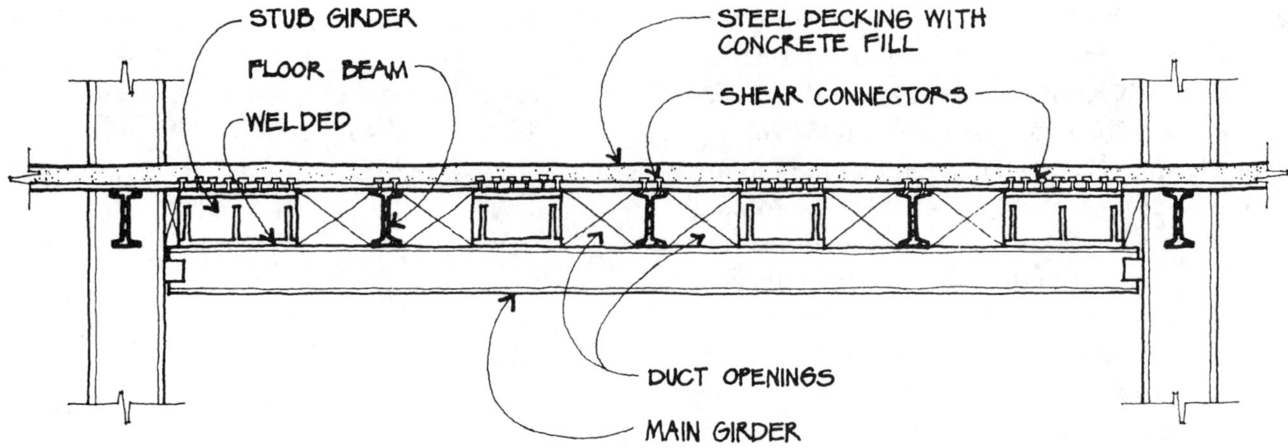

STUB GIRDER SYSTEM

concrete members and the spacing of reinforcing bars. In general, the maximum size of coarse aggregate should not exceed 1/5 of the narrowest dimension between sides of forms (A) or 3/4 of the clear spacing between reinforcing bars (B). Usually, more water is required for smaller size coarse aggregates than for larger maximum sizes (C). For a given water-cement ratio, then, the amount of cement required increases as the maximum size of coarse aggregate decreases. Therefore, for economy, the maximum size of coarse aggregate should be as large as possible. D is the incorrect statement and therefore the answer to this question.

276. C. The stub girder system consists of main steel girders framed between columns, above which short lengths of stub girders are welded. They are typically about five feet long and spaced about five feet apart. Shear connectors are welded to the tops of the stub girders to provide composite action with the concrete slab. Transverse to the stub girders are floor beams of the same depth, with shear connectors welded to their top flanges. The concrete slab

above the stub girders acts with the main girder to form a sort of Vierendeel truss, in which the main girder is the bottom chord, the slab is the top chord, and the stub girders are the verticals. Note that the concrete floor slab may be either a structural concrete slab or a composite steel deck with concrete fill. C is therefore correct.

277. C. This question tests your ability to use a table from the AISC Manual, which is on page 68, to select the most economical column section. We read down the left column to a length of 16 feet and then read across to the allowable axial load in kips for each column section under the heading 36 ($F_y =$ 36 ksi for ASTM A36 steel). For the W10 $\times$ 54, we read 251 kips; for the W10 $\times$ 49, we read 228 kips; for the W10 $\times$ 45, we read 180 kips; and for the W10 $\times$ 39, we read 154 kips. We select the W10 $\times$ 49 (choice C), as it is the lightest W10 column with a capacity of at least 200 kips.

278. B. In scheme B, simulated continuity is provided by hanging the center beam from the outer beams, thereby reducing the positive moment in the end spans (I).

In addition, the effective span of the center beam is reduced from L, the column spacing, to the distance between hinges, resulting in a smaller positive moment in the center span (II). Scheme B results in greater loads in the interior columns and smaller loads in the exterior columns (III is not correct). As I and II are advantages of scheme B, B is the correct answer.

279. D. If the upper soils are strong enough to support the building loads, then footings at a relatively shallow depth can be used. In this case, as the upper soils consist of loose fill, these loose soils must be removed and recompacted (III), if they are to support building loads. Alternatively, the footings could extend through these loose upper soils into the stronger soils below (II or IV), or the building loads could be spread over a large area (I), so that the average bearing pressure will be low and the system made rigid to bridge over localized areas of poor soil. Because all four systems might be appropriate, D is the correct answer.

280. A. The pressure applied to a retaining wall by the retained earth is usually assumed to vary uniformly from zero at the top to a maximum at the bottom, as shown in A and C. The soil pressure under the footing is maximum at the toe and minimum at the heel, as shown in both A and B. Only A has both correct diagrams, and is therefore the correct answer.

281. A. The axial load-carrying capacity of a steel column is determined by the strength of the steel used in the column and the tendency of the column to buckle. The buckling tendency is a function of the length of the column (1), its radius of gyration (r), and the relative fixity of its ends (K). In this case, the base plate with two bolts has very little resistance to rotation and can therefore be considered pinned. Because the column is part of a moment-resisting steel frame, its top is more or less fixed against rotation, but it can translate (move horizontally). The correct answer is therefore A.

282. D. Certain types of clay expand when wet and shrink when dried, and buildings supported on such expansive soils may suffer damage. To minimize or prevent damage, the foundations should be placed below the depth of seasonal moisture change, so that the moisture content of the subsoil remains relatively constant.

283. C. Choice A defines a gravity wall, B refers to a crib wall, C correctly defines a counterfort wall, and D describes a cantilever wall.

284. D. Exam questions often test candidates' understanding of how structures respond to load. In this case, the pressure of the retained earth against the wall varies from zero at the top to a maximum at the bottom, which causes the wall to cantilever off its footing. The soil pressure under the footing varies from a maximum under the toe to a minimum under the heel. The resulting distortions and stresses are shown below.

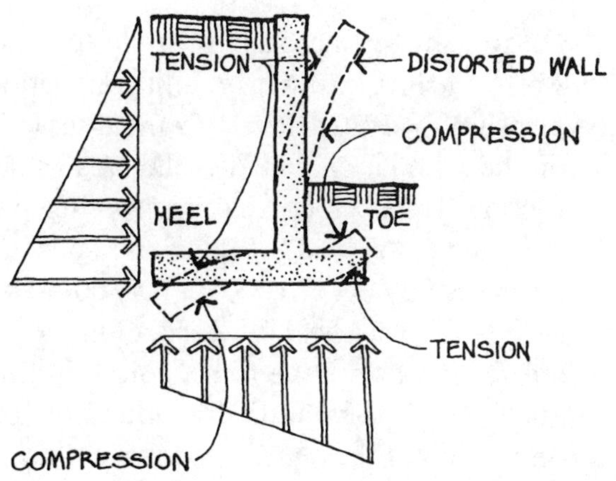

285. B. The retaining walls illustrated are all cantilever walls, which resist the lateral soil pressure by bending. The reinforcing steel must be placed in the tension side of the wall, which is the face closest to the retained earth. Thus, the wall reinforcing is correctly located in all of the illustrations, except B. C has reinforcing in both faces, which may be required in some situations. The bottom of the footing is in tension at the toe, while the top is in tension at the heel. Thus, all four cases show the footing reinforcing steel correctly.

286. C. Generally, rock is the best bearing material, followed by sands and gravels (I and II). Fine-grained soils, such as silts and clays, are usually adequate to support building foundations (III), but may require investigation. However, organic soils, such as peat (IV), are usually unacceptable for the support of buildings.

287. A. The pressure of the retained earth against a retaining wall varies linearly, from zero at the top to a maximum value at the bottom, as shown in A, B, and C. The bearing pressure under the base is maximum under the toe and minimum under the heel, as shown in A and D. Only A shows the correct pressure diagrams for both the stem and base.

288. D. Remember the basic formula for unit stress f = P/A? Similarly, unit foundation pressure f = P/A, where P is the total load on the foundation and A is the area of the footing. Transposing, required footing area A = foundation load P ÷ allowable soil bearing pressure f = (120,000# + 150,000#) ÷ 4,000#/ft.^{2} = 67.5 sq.ft. Most column pads are square, as are the four choices in this question. To determine the required side dimension of the square pad, we calculate $\sqrt{67.5}$ = 8.22 feet, and we therefore select answer D, 8'– 3" × 8' – 3".

289. C. The pressure varies linearly from zero at the top to a value at the bottom equal to 30 pounds per cubic foot times the depth of 10 feet, or 300 pounds per foot, per lineal foot of wall, as shown below. The total lateral force is equal to the area of the pressure diagram = 10 × 300/2 = 1,500#.

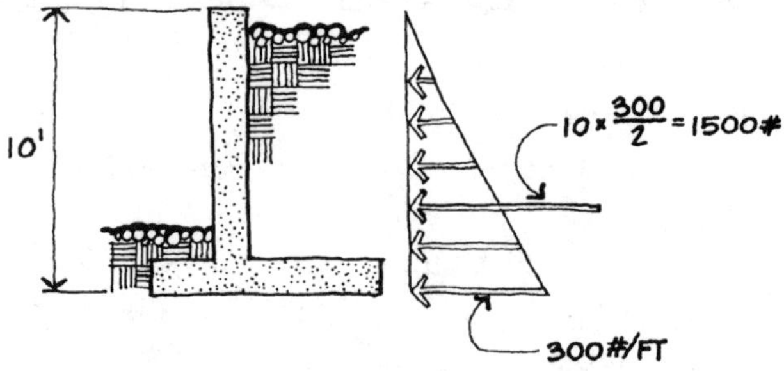

290. D. The resistance of a retaining wall to sliding is provided by friction between the footing and the underlying soil and by earth pressure in front of the toe. Where these are insufficient, additional sliding resistance may be obtained by constructing the footing with an integral key (A), or by making the footing wider (B) or deeper (C). Increasing the amount of reinforcing steel in the footing has no effect on sliding resistance. D is therefore the correct answer.

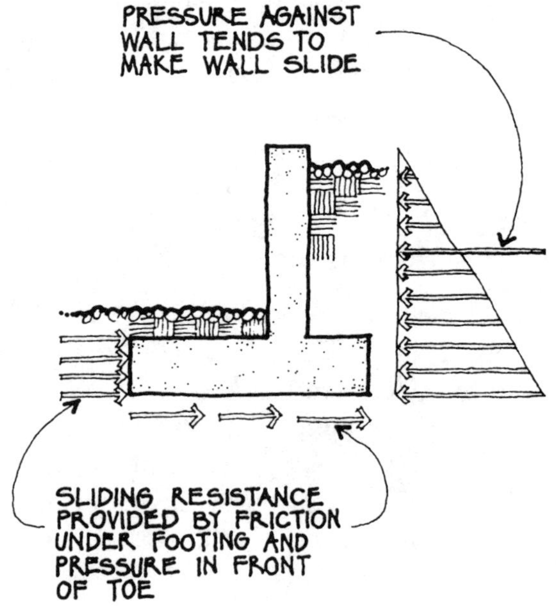

291. D. For concentrated columns loads, the most common type of footing is a square pad centered under the column, which results in a uniform soil pressure acting upward on the footing. However, if the column is located close to a property line, the footing cannot be centered under the column, and the resulting eccentricity between the center of the column and the center of the footing would cause the soil pressure distribution to be non-uniform. This could result in undesirable footing settlement, or an uneconomically large footing. To avoid these problems, a combined footing, as shown in the sketch, is often used, in which the resultant of the column loads coincides with the centroid of the combined footing. D is correct.

292. B. If the upper soils were strong enough to support the building loads, then footings at a shallow depth could be used. However, because the upper soils consist of loose fill, we must penetrate through the fill to bear on the dense sand below. Choices III and IV are therefore likely to be appropriate foundation systems (correct answer B). What about choices I and II? Wouldn't they also be appropriate? Not very likely; removal and recompaction of fill is usually economical up to a depth of about six feet, not 15 feet. And footings extending through the fill into the dense sand might be economical up to several feet in depth, but not 15 feet as in this question.